ASTRONOMY AND ASTROPHYSICS LIBRARY

More information about this series at http://www.springer.com/series/848

Thomas L. Wilson • Susanne Hüttemeister

Tools of Radio Astronomy – Problems and Solutions

Second Edition

Springer

Thomas L. Wilson
Max-Planck-Institut für Radioastronomie
Bonn, Germany

Susanne Hüttemeister
Zeiss-Planetarium Bochum
Bochum, Germany

ISSN 0941-7834   ISSN 2196-9698   (electronic)
Astronomy and Astrophysics Library
ISBN 978-3-319-90819-9   ISBN 978-3-319-90820-5   (eBook)
https://doi.org/10.1007/978-3-319-90820-5

Library of Congress Control Number: 2018942231

Cover illustration: In the foreground are dipole antennas of Long Wavelength Array station 1 (LWA1). In the background are some of the twenty-eight 25-meter paraboloids of the Jansky-Very Large Array (JVLA), of the National Radio Astronomy Observatory (NRAO). The LWA1 consists of 256 dual polarization, active dipole antennas (designed by the Naval Research Laboratory, NRL), operating between 10 MHz and 80 MHz. LWA1 was conceived by the NRL, designed and constructed in collaboration with university partners, and operated by the University of New Mexico. The JVLA is being equipped with receivers for operation at frequencies <1 GHz as part of a project involving NRL and the NRAO, and may eventually be able to operate in conjunction with LWA1. (Photo courtesy of J. Monkiewicz, School of Earth and Space Exploration, Arizona State University.)

This Springer imprint is published by the registered company Springer International Publishing AG
The registered company address is: Gewerbestrasse 11, 6330 Cham, Switzerland

Preface

This book of problems and solutions is an update of the first edition. The largest changes are (1) the addition of new problems, (2) the elimination of problems no longer relevant, and (3) a rearrangement to follow the exact order and numbering of the problem sets at the end of corresponding chapters in *Tools of Radio Astronomy*, 6th edition (hereafter 'Tools') by Wilson, Rohlfs and Hüttemeister (Springer-Verlag, 2013). We have tried to make this book as self-contained as possible. Thus, we have included figures, equations, and tables from 'Tools' that are directly relevant to the problems. In addition, we have added a few additional, necessary explanations in the formulation of the problems and the solutions. Usually, we have used a notation with lower case names for chapters and problem which are in this volume, but upper case when these are in 'Tools'.

In cases where the problems are more complex than usual, these have been divided into subsets. The problems themselves are of two types:

- Exercises which are a direct application of the material presented in the text. If use of specific equations in 'Tools' is needed, these are given.
- An extension of material or an alternative presentation of material in 'Tools'. This type of problem is identified by an asterisk (*). The number of such problems has been reduced to a minimum.

We have made use of a number of sources dealing with the interstellar medium, electromagnetic theory, and modern physics. Where applicable, we quote relevant original literature citations. We have not given general references for each chapter. These are to be found in 'Tools'.

In the following, we summarize the rationale for this book, taken from the preface of our first edition:

"We believe that a familiarity with orders of magnitude, typical estimates, and the basic understanding which one needs in observational radio astronomy can be learned only by practice; this means problem solving. Since there are only a few solved problems in 'Tools' itself, we decided to compose a set of ∼200 problems (many multi-part) which apply the principles set forth in 'Tools'. In addition, we wanted to give the flavor of the current state of radio astronomy, showing what is possible. We have tried to select examples from all branches of radio astronomy. This is done to achieve a 'practice-oriented' presentation which makes use of current instrumental parameters and our present-day understanding of source parameters. These problems can be considered as astronomical applications of the basic physics encountered at the level of final-year undergraduates."

A recurring problem is the choice of units. Astronomers prefer the CGS system, whereas practical work is greatly facilitated by using MKS units such as volts and amperes. We have tried to use the simplest approach in all situations. We give the most often used quantities in the following table. For a specific problem, we give the quantity needed for that problem in the problem itself.

The relation of CGS and other systems of units can be found in the website: *NRL Plasma Formulary*.

Bonn, Germany   Thomas L. Wilson
Bochum, Germany   Susanne Hüttemeister
February 2018

Some Relevant Physical Constants in CGS Units

Quantity	Symbol	Value
Velocity of light	c	2.99792×10^{10} cm s^{-1}
Gravitational constant	G	6.67×10^{-8} dyne cm^2 g^{-1}
Planck's constant	h	6.626×10^{-27} erg s
Charge of the electron	e	4.80325×10^{-10} electrostatic units (esu)
Mass of the electron	m_e	9.11×10^{-28} g
Mass of the proton	m_p	1.672×10^{-24} g
Boltzmann's constant	k	1.380×10^{-16} erg degree^{-1}
Avogadro's number	N_A	6.022×10^{23} mole^{-1}
One electron volt	eV	1.602×10^{-12} erg
Stefan–Boltzmann constant	σ	1.80×10^{-5} erg cm^{-2} degree^{-4} s^{-1}
Bohr radius	a_0	5.29×10^{-9} cm
Debye	D	2.53×10^{-18} esu cm

Some Relevant Astronomical Constants

Quantity	Symbol	Value
Astronomical unit	AU	1.45979×10^{13} cm
Parsec	pc	3.085678×10^{18} cm
Light year	lt yr	9.460530×10^{17} cm
Mass of the Sun	$M_\odot$	1.989×10^{33} g
Radius of the Sun	$R_\odot$	6.9599×10^{10} cm
Luminosity of the Sun	$L_\odot$	3.826×10^{33} erg s^{-1}
Mass of the earth	M_e	5.976×10^{27} g
Radius of the Earth (equator)	R_e	6378.164 km
Radius of Jupiter	R_J	71.492 km
Mass of the Galaxy	$M_{\text{Milky Way}}$	$\sim 10^{11} M_\odot$

Contents

Chapter 1
Radio Astronomical Fundamentals

1. If the average electron density in the interstellar medium (ISM) is $0.03\,\mathrm{cm}^{-3}$, what is the lowest frequency of electromagnetic radiation which one can receive due to the effect of this plasma? Compare this to the ionospheric plasma cutoff frequency if the electron density, N_e, in the ionosphere is $\sim 10^5\,\mathrm{cm}^{-3}$. Use the equation on page 4 of 'Tools', which is: $\nu_p = 8.97 \times (0.03)^{0.5}$.

2. (a) A researcher measures radio emission at a frequency of 250 kHz and finds that the emission is present over the whole sky with a brightness temperature of 250 K. Could the origin of this radiation be the earth's ionosphere?
(b) Assume that the source fills the entire visible sky, taken to be a half hemisphere. What is the power received by an antenna with $A = 1\,\mathrm{m}^2$ collecting area in a $B = 1\,\mathrm{kHz}$ bandwidth?

3. The downward-pointing radar satellite, *Cloudsat*, is moving in a polar orbit at an altitude of 500 km. The operating frequency is 94 GHz. Assume that the power is radiated over a hemisphere. The peak power will be 1500 W, uniformly distributed over a bandwidth of 1 GHz. If no power is absorbed in the earth's atmosphere, what is the peak flux density of this satellite when it is directly overhead? This radar is transmitting 3% of the time (duty cycle). What is the *average* power radiated and the corresponding flux density?

4. A unit commonly used in astronomy is flux density, S_ν, the Jansky (Jy). One Jy is $10^{-26}\,\mathrm{W\,m}^{-2}\,\mathrm{Hz}^{-1}$. Calculate the flux density, in Jy, of a microwave oven with an output of 1 kW at a distance of 10 m if the power is radiated over all angles and is uniformly emitted over a bandwidth of 1 MHz.

5. (a) What is the flux density, S_ν, of a source which radiates a power of 1 kW in the microwave frequency band uniformly from 2.6 to 2.9 GHz, when placed at the distance of the Moon $(3.84 \times 10^5\,\mathrm{km})$? Repeat for an identical source if the radiation is in the optical frequency band, from 3×10^{14} to $8 \times 10^{14}\,\mathrm{Hz}$.

© Springer International Publishing AG, part of Springer Nature 2018
T. L. Wilson, S. Hüttemeister, *Tools of Radio Astronomy – Problems
and Solutions*, Astronomy and Astrophysics Library,
https://doi.org/10.1007/978-3-319-90820-5_1

(b) If we *assume* that the number of photons is uniform over the band, what is the average energy, $E = h\nu$, of a photon? Use this average photon energy and the power to determine N, the number of photons. How many photons pass through a $1\,m^2$ area in one second in the optical and radio frequency bands?

6. In the near future there may be an anti-collision radar installed on automobiles. It will operate at $78.5\,GHz$. If the bandwidth is $10\,MHz$, and at a distance of $3\,m$, the power per area is $10^{-9}\,W\,m^{-2}$. Assume the power level is uniform over the entire bandwidth of $10\,MHz$. What is the flux density of this radar at $1\,km$ distance? A typical large radio telescope can measure to the mJy ($=10^{-29}\,W\,m^{-2}\,Hz^{-1}$) level. At what distance will such radars disturb such radio astronomy measurements?

7. If the intensity of the Sun peaks in the optical range, at a frequency of about $3.4 \times 10^{14}\,Hz$, what is the temperature of the Sun? Use the Wien displacement law Eq. (1.25) of 'Tools', which is: $\left(\frac{\nu_{max}}{GHz} = 58.789\,\frac{T}{K}\right)$. If all of the power is emitted only between 3 and $4 \times 10^{14}\,Hz$, how many photons per cm^2 arrive at the earth when the Sun is directly overhead? What is the power received on earth per cm^2? A value for the solar power is $135\,mW$ per cm^2. How does this compare to your calculation?

Problems number 8 and 9 were shifted to a later chapter, so the next problem in this chapter is number 10.

10. Show that Eq. (1.34), which is,

$$\left(\frac{S_\nu}{Jy}\right) = 2.65\,T_B \left(\frac{\theta}{arcmin}\right)^2 \left(\frac{\lambda}{cm}\right)^{-2}$$

can be obtained from Eq. (1.33), which is, $S_\nu = \frac{2k\nu^2}{c^2} T_B\,\Delta\Omega$ Extend the relation to arc seconds, wavelength in millimeters and milli Jy, to obtain:

$$\left(\frac{S_\nu}{mJy}\right) = 73.6\,T_B \left(\frac{\theta}{arcsec}\right)^2 \left(\frac{\lambda}{mm}\right)^{-2}$$

11. Suppose the maximum observed temperature of the galaxy is $10^5\,K$ at $14.6\,m$ wavelength. If this is measured with a $24°$ gaussian beam, what is the flux density contained within the telescope beam? If the bandwidth used is $100\,kHz$ and the collecting area is $800\,m^2$, what is the received power?

12. The discrete source Cassiopeia A has an observed peak brightness temperature of $7 \times 10^4\,K$ at $20\,MHz$. If the antenna has a beamsize of $24°$, what is the flux density? The source has a size of $4\,arcmin$. Using the approximate relation

$$T_{source} \cdot \theta_{source}^2 = T_{observed} \cdot \theta_{beam}^2$$

What is the actual source temperature, T_{source}? This is proven mathematically in problem 6 of chapter 8.

13*. In Leighton ('Principles of Modern Physics', 1959, Appendix B, p. 725), the units for the Planck function, B, are mixed, with the wavelength, λ, in micrometers, μm, power in Watts, area in centimeters and angles in steradians. Show that when physical constants are inserted in Eq. (1.22) from 'Tools':

$$B_\lambda(T) = \frac{2hc^2}{\lambda^5} \frac{1}{e^{hc/k\lambda T} - 1}$$

the result is:

$$\left(\frac{B_\lambda(T)}{\mathrm{W\ cm^{-2}\ \mu^{-1}\ sterad^{-1}}} \right) = 1.19 \cdot 10^4 \left[\lambda(\mu m) \right]^{-5} \frac{1}{e^y - 1}$$

where $y = 1.44 \cdot 10^4 / \lambda(\mu m)\, T$. Another form of this relation, that could be used for infrared spectroscopy, is

$$\int \left(\frac{B_\nu(T)}{\mathrm{ergs\ s^{-1}\ cm^{-2}\ sterad^{-1}}} \right) = 4.9 \cdot 10^{-17} \left[\nu(\mathrm{GHz}) \right]^4 \, \Delta V(\mathrm{km\ s^{-1}}) \frac{1}{e^x - 1}$$

where $x = 4.84 \cdot 10^{-2}\, \nu(\mathrm{GHz})/T$. Show that in the Rayleigh-Jeans limit, this takes on the simpler form:

$$\int \left(\frac{B_\nu(T)}{\mathrm{ergs\ s^{-1}\ cm^{-2}\ sterad^{-1}}} \right) = 10^{-15}\, (\nu(\mathrm{GHz}))^3 \, T\, \Delta V(\mathrm{km\ s^{-1}})$$

14. Insert values of physical constants in Eq. (1.13) of 'Tools', which is,

$$B_\nu(T) = \frac{2h\nu^3}{c^2} \frac{1}{e^{h\nu/kT} - 1}$$

to obtain:

$$\left(\frac{B_\nu}{\mathrm{Jy\ sterad^{-1}}} \right) = 1.47 \times 10^3\, \nu(\mathrm{GHz})^3 \frac{1}{e^x - 1}$$

where $x = h\nu/kT = 4.81 \cdot 10^{-2}\, \nu(\mathrm{GHz})/T$.

15*. If Jupiter has $T_B = 150\,\mathrm{K}$, with $\theta \approx 40''$, what is S_ν at 1.4 GHz? At 115 GHz? Repeat for the HII region Orion A, with $\theta = 2.5'$, with $T_B = 330\,\mathrm{K}$ at 4.8 GHz, and $T_B = 24\,\mathrm{K}$ at 23 GHz. Use the result of problem 10 (also Eq. (1.33) of 'Tools') of this chapter.

16. Show that the solid angle in steradians for a Jupiter-like planet (in an extrasolar system) with a radius in terms of R_J (where R_J=71,492 km) at a distance D, in parsecs, is:

$$\Delta\Omega = 1.7 \cdot 10^{-17} \cdot \left(\frac{R(R_J)}{D(pc)}\right)^2$$

Using the result given in problem 10 show that the flux density relation is:

$$S_\nu(Jy) = 2.6 \cdot 10^{-14} \, \nu(GHz)^3 \left(\frac{1}{e^x - 1}\right) \cdot \left(\frac{R(R_J)}{D(pc)}\right)^2$$

where x is

$$x = 4.81 \cdot 10^{-2} \, \nu(GHz)/T$$

Estimate the detectability of a "Hot Jupiter", that is, a Jupiter in the formation process, using Eq. (1.1), if such an object has T=2500 K, R=30 R_J (see Wolf and D'Angelo 2005 Ap. J. 619, 1114). For D=20 parsecs and ν=345 GHz or $\lambda = 0.87$ mm, show that S_ν=36 μJy.

17. Show that the flux density of a solar-like star with a radius $R_\odot$ is given by:

$$S_\nu(Jy) = 2.5 \cdot 10^{-12} \, \nu(GHz)^3 \, \frac{1}{e^x - 1} \cdot \left(\frac{R(R_\odot)}{D(pc)}\right)^2$$

where x is as given in the last problem. Show that if we use $R = R_\odot$, $D = 20$ pc and T=5700 K, the flux density at $\lambda = 0.8$ mm is $S_\nu = 35$ mJy. If, for an O star, these values are T=4.1 10^5 K, $R = 18.2 \, R_\odot$, and if $S_\nu(Jy) = 10^{-3}$, for a detection, what is this distance in pc?

18. For Fig. 1.6 from 'Tools' (here figure 1.1), determine the frequency that corresponds to the peak of 150 K radiation from Jupiter. At three times this frequency, determine the intensity of the Rayleigh-Jeans and Planck expressions. For one tenth of the peak frequency, determine these values and the prediction from the Wien Law (see problem 7). Often the Planck temperature scale is used in radio studies of the planets. The 2.73 background radiation follows the Planck relation. Show that the peak occurs at 160 GHz. Use Eq. (1.32) from 'Tools', which is $T_B = \frac{c^2}{2k\,\nu^2} I_\nu \Delta\Omega$, to determine the temperature in the Rayleigh-Jeans approximation at 115 GHz. [Hint: use $\delta\Omega = 4\pi$]

19. A cable has an optical depth, τ, of 0.1 and a temperature of T = 300 K. A signal of peak temperature $T_b(0)$=1 K is connected to the input of this cable. Use Eq. (1.37) in 'Tools', which is $T_b(s) = T_b(0) + T(1 - e^{-\tau})$ to analyze this situation. What is $T_b(s)$, the temperature of the output of the cable? Repeat the problem for T = 100 K.

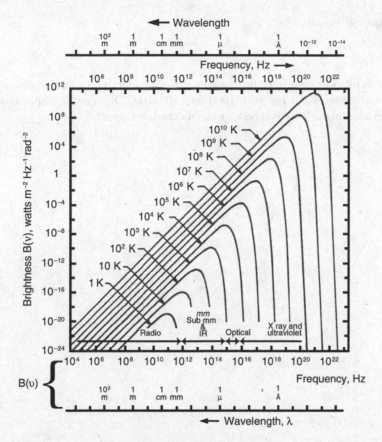

Fig. 1.1 This is Fig. 1.6 in 'Tools'. This is a plot of Planck spectra for black bodies of different temperatures versus wavelength and frequency (for problem 18)

What is the signal-to-noise ratio for these two cases, using signal = 1 K, and noise from the cable contribution?

20. A signal passes through two cables with the same optical depth, τ. These have temperatures T_1 and T_2, with $T_1 < T_2$. Which cable should be connected first to obtain the lowest output power from this arrangement?

21. Eq. (1.43) of 'Tools', namely, $P = k T \, \Delta \nu$, describes the power radiated in one dimension. If a microwave oscillator delivers 1 mW of power uniformly over a bandwidth of 1 Hz, what is the equivalent temperature T? Since the physical temperature of such an oscillator is ~ 300 K, this is an example of a *non-thermal* process.

22*. Eq. (1.40) is given by:

$$T_{b_0} = (1 - r)T_0 + rT_s$$

Using this relation, determine the noise contribution of a room temperature reflector with an r value of 0.9 for 100, 1000 and 10^4 GHz. Discuss the implication for infrared telescopes. Why is it better to cool such telescopes?

Chapter 2
Electromagnetic Wave Propagation Fundamentals

1. There is a proposal to transmit messages to mobile telephones in large U.S. cities from a transmitter suspended below a balloon at an altitude of 40 km. Suppose the city in question has a diameter of 40 km. What is the solid angle to be illuminated? Suppose mobile telephones require an electric field strength, E, of 200 μV per meter. If one uses $S = E^2/R$ with $R = 50\,\Omega$, what is the E field at the transmitter? How much power must be transmitted? At what distance from the transmitter would the microwave radiation reach the danger level, $10\,\text{mW}\,\text{cm}^{-2}$?

2. Radiation from an astronomical source at a distance of 1.88 kpc, (1 pc=3.08 × 10^{18} cm) has a flux density of 10^3 Jy over a frequency band of 600 Hz. If it is isotropic, what is the power radiated? Suppose the source size is 1 milli arc second (see Eq. (1.34) given in problem 10 of chapter 1). What is the value of T_B?

3. Suppose that $v_\text{phase} = \dfrac{c}{\sqrt{1-(\lambda_0/\lambda_c)^2}}$ and that $v_\text{phase} \times v_\text{group} = c^2$ What is v_group? Evaluate both of these quantities for $\lambda_0 = \frac{1}{2}\lambda_c$.

4. There is a 1 D wave packet. At time t=0, the amplitudes are distributed as $a(k) = a_0 \exp(-k^2/(\Delta k)^2)$, where a_0 and Δk are constant. From the use of Fourier transform relations in Appendix A, determine the product of the width of the wave packet, Δk, and the width in time, Δt.

5. Repeat problem 4 with $a(k) = a_0 \exp(-(k - k_0)^2/(\Delta k)^2)$. Repeat for $a(k) = a_0$ for $k_1 < k < k_2$, otherwise $a(k) = 0$.

6. Assume that pulsars emit narrow periodic pulses at all frequencies simultaneously. Differentiate Eq. (2.67), which is,

$$\tau_\text{D} = \frac{L}{c} + \frac{e^2}{2\pi c\, m_e} \frac{1}{v^2} \int\limits_0^L N(l)\,dl$$

© Springer International Publishing AG, part of Springer Nature 2018
T. L. Wilson, S. Hüttemeister, *Tools of Radio Astronomy – Problems and Solutions*, Astronomy and Astrophysics Library,
https://doi.org/10.1007/978-3-319-90820-5_2

to show that a narrow pulse (width of order $\sim 10^{-6}$ s) will traverse the radio spectrum at a rate, in MHz s^{-1}, of $\dot{\nu} = 1.2 \times 10^{-4}$ (DM)$^{-1}$ ν [MHz]^3.

7. (a) Show that using a receiver bandwidth B will lead to the smearing of a very narrow pulse, which passes through the ISM with dispersion measure DM, to a width $\Delta t = 8.3 \times 10^3$ DM $[\nu$ (MHz)$]^{-3}$ B s.
(b) Show that the ionosphere (electron density 10^5 cm^{-3}, height 20 km) has little influence on the pulse shape at 100 MHz.

8. (a) Show that the smearing of a short pulse, Δt, in milli seconds per MHz of receiver bandwidth, is $(202/\nu_{\text{MHz}})^3$ DM.
(b) If a pulsar is at a distance of 5 kpc, and the average electron density is 0.05 cm^{-3}, find the smearing at 400 MHz. Repeat for 800 MHz.

9. Suppose you would like to detect a pulsar located at the center of our Galaxy. The pulsar may be behind a cloud of ionized gas of size 10 pc, and electron density 10^3 cm^{-3}. Calculate the dispersion measure, DM. What is the bandwidth limit if the observing frequency is 1 GHz, and the pulsar frequency is 30 Hz?

10. A typical value for DM is 30 cm^{-3} pc, which is equivalent to an electron column density of 10^{20} cm^{-2}. For frequencies of 400 MHz and 1000 MHz, use Eq. (2.71), which is,

$$\frac{\Delta \tau_D}{\mu s} = 4.148 \times 10^9 \left[\frac{DM}{\text{cm}^{-3} \text{ pc}} \right] \left[\frac{1}{\left(\frac{\nu_1}{\text{MHz}}\right)^2} - \frac{1}{\left(\frac{\nu_2}{\text{MHz}}\right)^2} \right]$$

Use this to predict how much a pulse will be delayed relative to a pulse at an infinitely high frequency.

11. To resolve a pulse feature with a width of 0.1 μs at a received frequency of 1000 MHz and DM = 30 cm^{-3} pc, what is the maximum receiver bandwidth? For this, one needs Eq. (2.73): $b_{\text{MHz}} = 1.205 \times 10^{-4} \frac{1}{\text{DM}} \nu^3$(MHz) $\times \tau_s$, where b is the bandwidth.

12. In Fig. 2.2 of 'Tools', reproduced here as fig. 2.1, at about the 10% level, the pulse width is 10 ms. What bandwidth is needed to resolve this pulse for a DM=295 cm^{-3} pc?

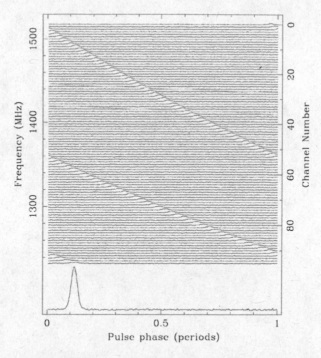

Fig. 2.1 This is Fig. 2.2 in 'Tools'. It is referred to in problem 12. The diagram shows pulse arrival times at frequencies from 1.24 to 1.5 GHz. The source is the pulsar B1356-60. The single pulse shown at the bottom has been coherently added from single pulses. It has a period of 128 ms. The Dispersion Measure, DM, is 295 cm^{-3} pc

Chapter 3
Wave Polarization

1. In optical astronomy, the convention for circular polarization is *opposite* to that used in radio astronomy, as used in this chapter. The difference is contained in the convention of the direction in which the wave is rotating. In the optical case, right-handed circular polarization has $\sin \delta < 0$. Make use of this change to interpret the sense of rotation of right-handed circular polarization (contrast with the description in the caption of Fig. 3.3 of 'Tools', given here as figure 3.1).

For this and the following problems, you may need the Stokes parameters, as given in Eq. (3.39) of 'Tools'. These are:

$$S_0 = I = E_1^2 + E_2^2$$
$$S_1 = Q = E_1^2 - E_2^2$$
$$S_2 = U = 2E_1 E_2 \cos \delta$$
$$S_3 = V = 2E_1 E_2 \sin \delta$$

2. A source is 100% linearly polarized in the north–south direction. Express this in terms of Stokes parameters. Use the equations given in the statement of problem 1, above.

3. If the degree of polarization is 10% in Eq. (3.55), namely,

$$p = \frac{\sqrt{S_1^2 + S_2^2 + S_3^3}}{S_0}$$

© Springer International Publishing AG, part of Springer Nature 2018
T. L. Wilson, S. Hüttemeister, *Tools of Radio Astronomy – Problems and Solutions*, Astronomy and Astrophysics Library,
https://doi.org/10.1007/978-3-319-90820-5_3

with $S_3=0$, $S_1=S_2$ in Eq. (3.53). In another form, in contrast to problem 2 of this chapter, this is,

$$S_0 = I = I(0°, 0) + I(90°, 0)$$
$$S_1 = Q = I(0°, 0) - I(90°, 0)$$
$$S_2 = U = I(45°, 0) - I(135°, 0)$$
$$S_3 = V = I(45°, \tfrac{\pi}{2}) - I(135°, \tfrac{\pi}{2})$$

what is the state of polarization?

4. Intense spectral line emission at 18 cm wavelength is caused by maser action of the OH molecule. At certain frequencies, such emission shows nearly 100% left circular polarization, but no linear polarization. Express this in terms of Stokes parameters.

5. In Fig. 3.3 of 'Tools', given here as figure 3.1, if we reverse the radio astronomical convention, so that if a wave is receeding from us, it is counterclockwise, what is the sense of polarization? Is the wave right-hand circular or left-hand circular? Note that this is the optical definition, opposite to that in radio astronomy.

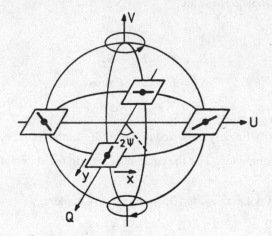

Fig. 3.1 This is Fig. 3.3 in 'Tools'. It shows Stokes parameters for polarization and the Poincaré sphere. Taking the angles 2ψ and 2χ as angles in a polar coordinate system, each point on the surface of the resulting sphere corresponds to a unique state of polarization. The positions on the equator ($2\chi = 0$) correspond to *linear polarization*, those at the northern latitudes ($2\chi > 0$) contain *right-handed* circular polarization, while those on the southern hemisphere contain *left-handed*. If we orient the (x, y) coordinate system parallel to Q and U, the linear polarization of the waves are oriented as indicated. A simple rule-of-thumb is that for an approaching wave counterclockwise is right-hand circular polarization. This is the radio astronomy convention. Note that this definition is the opposite of that used in the optical

6. Determine the *upper limit* of the angle through which a linearly polarized electromagnetic wave is rotated when it traverses the ionosphere. Use Eq. (3.70) in 'Tools', which is: $\frac{\Delta\psi}{\text{rad}} = 8.1 \times 10^5 \left(\frac{\lambda}{m}\right)^2 \int_0^{L/pc} \left(\frac{B_\parallel}{\text{Gauss}}\right) \left(\frac{N_e}{\text{cm}^{-3}}\right) d\left(\frac{z}{pc}\right)$. First, one must calculate the rotation measure, RM. This is given Eq. (3.70).

(a) Find RM using Eq. (3.71) of 'Tools'): $\frac{\text{RM}}{\text{rad}} = 8.1 \times 10^5 \int_0^{L/pc} \left(\frac{B_\parallel}{\text{Gauss}}\right) \left(\frac{N_e}{\text{cm}^{-3}}\right)$ $d\left(\frac{z}{pc}\right)$ with the following parameters: an ionospheric depth of 200 km, an average electron density of 10^5 cm^{-3} and a magnetic field strength (assumed to be parallel to the direction of wave propagation) of 1 G.

(b) Carry out the calculation for the Faraday rotation, $\Delta\psi$, for frequencies of 100 MHz, 1 and 10 GHz.

(c) What is the effect if the magnetic field direction is perpendicular to the direction of propagation? What is the effect on circularly polarized electromagnetic waves?

(d) Repeat for the conditions which hold in the solar system: the average charged particle density in the solar system is 5 cm^{-3}, the magnetic field 5 μG and the average path 10 AU (=1.46 × 10^{14} cm). What is the maximum amount of Faraday rotation of an electromagnetic wave of frequency 100 MHz, 1 GHz? Must radio astronomical results be corrected for this?

7. A 100% linearly polarized interstellar source is 3 kpc away. The average electron density in the direction of this source is 0.03 cm^{-3}. The magnetic field along the line-of-sight direction, $B_\parallel$, is 3 μG. What is the change in the angle of polarization at 100 MHz, at 1 GHz?

8. A right hand circularly polarized electromagnetic wave is sent perpendicular to a perfectly conducting metallic flat surface. The electromagnetic energy must be zero inside this conductor.

(a) Use a qualitative argument to show that the sense of the polarization of the reflected wave is opposite to that of the incoming wave.

(b) What is the effect of reflection on a linearly polarized signal?

9. If the DM for a given pulsar is 50, and the value of RM is 1.2×10^2, what is the value of the *average* line-of-sight magnetic field? If the magnetic field perpendicular to the line of sight has the same strength, what is the total magnetic field.

10. Consider a quasi-monochromatic wave with $\Delta\nu/\bar{\nu} = 0.1$ and $\nu = \nu_0$, a constant. Use Eq. (3.41), given here as:

$$V^{(r)}(t) = \int_0^\infty a(\nu)\cos[\phi(\nu) - 2\pi\nu t]\, d\nu.$$

with a(ν)=a_0, a constant, and $\phi(\bar{\nu}+\mu) = \phi_0$ likewise a constant. With these values, calculate A(t). This is an idealization, however is a commonly used approximation to describe wide band signals limited by narrow filters.

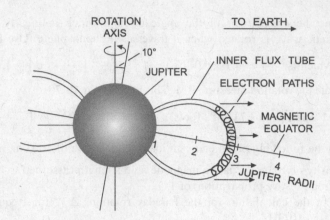

Fig. 3.2 The direction of the rotation axis and the direction of rotation of Jupiter are shown. The direction of the magnetic field, **B**, is shown as *thin lines*

11. Repeat problem 5, chapter 2, for the function

$$a(v) = a_0 e^{\left(-\frac{(v-v_0)}{\Delta v}\right)^2}$$

Show that $\Delta v\, \Delta t = 1$.

12. The planet Jupiter has a dipole magnetic field, **B** as shown in Fig. 3.2. Electrons trapped in this field move in circular orbits perpendicular to **B**. Describe qualitatively the polarization measured with a radio telescope beam which is small compared to all dimensions.

Chapter 4
Signal Processing and Receivers: Theory

1. The Gaussian probability distribution function with mean m is

$$p(x) = \frac{1}{\sigma\sqrt{2\pi}} e^{-(x-m)^2/2\sigma^2} .$$

(a) Show that $\int_{-\infty}^{+\infty} p(x)\,dx = 1$. If the first moment, or mean value m, is

$$m = \langle x \rangle = \int_{-\infty}^{+\infty} x p(x)\,dx$$

and the second moment is

$$\langle x^2 \rangle = \int_{-\infty}^{+\infty} x^2 p(x)\,dx ,$$

(b) find m and σ, the RMS standard deviation, where $\sigma = \langle x^2 \rangle - \langle x \rangle^2$. The third and fourth moments are defined in analogy with the definitions above. Determine the third and fourth moments of the Gaussian distribution.
(c) The relation between $\langle x^2 \rangle$ and $\langle x^4 \rangle$ has been used to study the noise statistics for very intense narrow band emission from an astronomical source at 18 cm (see Evans et al. 1972 Phys. Rev. A6, 1643; in addition this method has been used to eliminate interference encountered in passive satellite measurements, that is, 'kurtosis'). If the noise input has zero mean, and if the voltages $\langle v^2 \rangle$ and $\langle v^4 \rangle$ are compared, what would you expect the relation to be for a Gaussian distribution of noise?

2. For an input

$$v(t) = A \sin 2\pi \nu t$$

© Springer International Publishing AG, part of Springer Nature 2018
T. L. Wilson, S. Hüttemeister, *Tools of Radio Astronomy – Problems
and Solutions*, Astronomy and Astrophysics Library,
https://doi.org/10.1007/978-3-319-90820-5_4

calculate the Fourier Transform (FT), autocorrelation function and power spectrum. Note that this function extends to negative times. Repeat the calculation for

$$v(t) = A \cos 2\pi \nu t .$$

3. Calculate the power spectrum, S_ν, for the sampling function $v(t) = A$ for $-\tau/2 < t < \tau/2$, otherwise $v(t) = 0$, by taking the Fourier transform to obtain $V(\nu)$ and then squaring this. Next, form the autocorrelation of this function, and then FT to determine the power spectrum. Show that these two methods are equivalent.

4. Repeat the analysis in Problem 3, but shifting this function by a time $+\tau/2$: that is, $v(t) = A$ for $0 < t < \tau$, otherwise $v(t) = 0$. The FT shift theorem is given in Eq. (B5) in Appendix A:

$$f(x - a) \leftrightarrow e^{-i2\pi a s} F(s) .$$

Show that the result of this problem can be obtained from the result of Problem 3 by applying the shift theorem. What is the value of the shift constant, a?

5. Convolve the function $f_1(t)$, on the right side of figure 4.1 (also Fig. 4.10 in 'Tools') with itself, graphically. Show four steps in the relative offset τ and use symmetry to complete the calculation.

6. This problem refers to figure 4.2 (this is Fig. 4.5 in 'Tools'). Convolve the 'picket fence' function, in part (c) of that Figure with the function in part (a) graphically, in the frequency domain. That is, convolve part (b) with part (d). This is the mathematical representation of sampling in the time domain. If the adjacent samples overlap in the frequency domain, the results will be *aliased*. Show graphically that aliasing occurs if the time sampling rate is halved, i.e. in frequency from $\nu_0 = \frac{1}{T_0}$ to $\nu_0 = \frac{1}{2T_0}$.

7. Repeat problem 4 for the function $v(t) = A$ for $\tau < t < 2\tau$, and $-2\tau < t < -\tau$, otherwise $v(t) = 0$. The result can be interpreted as the frequency distribution calculated in Problem 5, modulated by $\cos 2\pi \nu \tau$. This is an example of the modulation property of Fourier transforms, in Appendix A, under 'Modulation', namely,

$$f(x) \cos x = \frac{1}{2}F(s - \nu) + \frac{1}{2}F(s + \nu) .$$

Fig. 4.1 This is Fig. 4.10 in 'Tools'. The shape on the right, $f_1(t)$, is used for the graphical convolution in problem 5

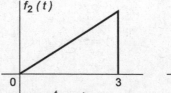
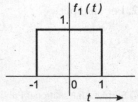

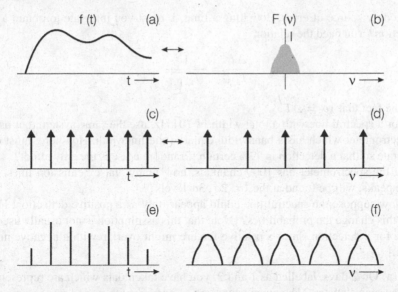

Fig. 4.2 This is for problem 6. It is also Fig. 4.5 in 'Tools'.. This shows the time and frequency distribution of sampled functions f(t): (**a**) The time variation, (**b**), the frequency behavior, (**c**) the time behavior of a regularly spaced sampling function (referred to as a "picket fence" function), (**d**) the frequency behavior of the "picket fence" function, (**e**) the time behavior of the sampled function, and (**f**) the frequency behavior of the function sampled with a "picket fence". The result in (**f**) is low pass filtered to recover the input as shown in **b**. As can be seen from these plots, the maximum frequency extent in (**b**) is smaller than the sampling rate, as shown in (**d**)

Table 4.1 Gaussian integrals used to determine noise statistics

σ	Value outside the curve	Value inside
1	0.3174	0.6826
2	0.0456	0.9544
3	0.0026	0.9974
4	0.0020	0.9980

8. Consider another aspect of the situation described in the last problem. We have a function $\cos(2\pi v_c t)\cos(2\pi v_s t)$, where $v_s = v_c + \Delta$, where $\Delta \ll v_c$. Apply the identity $\cos(x + y) = (1/2)[\cos(x + y) + \cos(x - y)]$. Check whether the modulation property of the Fourier transform applies.

9. Table 4.1 is a list of Gaussian integrals to determine the area within the boundary of the curve at the σ, 2σ, 3σ and 4σ levels is given in Table 4.1 here: (This is Table 4.2 in 'Tools').
(**a**) If you want to determine whether a feature is *not* noise at the 1% level, how many standard deviations from the mean must this signal be?
(**b**) Suppose you want to detect a continuum signal of peak temperature 10^{-2} K with a total power receiver with a system noise of $T_{sys}=200$ K, and a bandwidth, Δv of 500 MHz. Assume that this system is perfectly stable, that is random noise

is the only source of error. How long a time, τ must you integrate to obtain a 3σ detection? You need the relation

$$\frac{\Delta T}{T_{sys}} = \frac{1}{\sqrt{\Delta \nu \, \tau}}$$

and the fact that $1\sigma = \Delta T$.

(c) For a spectral line with a total width of 10 kHz, use the same system, but using a spectrometer which has a bandwidth equal to the linewidth. How long must one integrate so that a detection is 99% certain if random noise is the only effect?

(d) If the spectrometer has 1000 channels, how many "false" emission lines, i.e. noise peaks, will be found at the 1σ, 2σ, 3σ levels?

(e) Now suppose the spectral line could appear only as a positive deflection. How does this change the probabilities? [Note that this assumption is not usually used to argue for a detection, since a positive feature might overlap with a negative noise spike.]

10*. (a) On 2 days, labelled as 1 and 2, you have taken data which are represented by Gaussian statistics. The mean values are x_1 and x_2, with σ_1 and σ_2. Assume that the average is given by $\bar{x} = f x_1 + (1 - f)x_2$ and the corresponding $\sigma^2 = f^2\sigma_1^2 + (1 - f)^2\sigma_2^2$. Determine the value of f which gives the smallest $\bar{\sigma}$ by differentiating the relation for σ and setting the result equal to zero. Show that

$$\bar{x} = \left(\frac{\sigma_2^2}{\sigma_1^2 + \sigma_2^2}\right) x_1 + \left(\frac{\sigma_1^2}{\sigma_1^2 + \sigma_2^2}\right) x_2$$

and

$$\bar{\sigma}^2 = \left(\frac{\sigma_2^4}{(\sigma_1^2 + \sigma_2^2)^2}\right) \sigma_1^2 + \left(\frac{\sigma_1^4}{(\sigma_1^2 + \sigma_2^2)^2}\right) \sigma_2^2 .$$

(b) Use the relation $\sigma^2 \sim 1/(\text{time})$ to show that the expression for $\bar{x}$ reduces to the result, $\bar{x} = (1/(t_1 + t_2)) (t_1 x_1 + t_2 x_2)$.

11. Obtain Eq. (4.36) in 'Tools', given here:

$$F = \frac{S_1/N_1}{S_2/N_2} = \frac{N_2}{G N_1} = 1 + \frac{T_R}{T_1}$$

these variables are shown as figure 4.3 (and in Fig. 4.6 of 'Tools'). Justify the definition of the noise factor F in Eq. (4.36) based on the case of a *noiseless* receiver, i.e. one with $F = 1$. Show that this definition is consistent with the definition of receiver noise temperature

$$T_R = (F - 1) \times 290 \, K$$

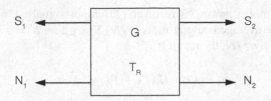

Fig. 4.3 A schematic diagram of a two port system needed for problem 12; this is also Fig. 4.6 of 'Tools'. The receiver is represented as a box, with the input signal S_1, and noise, N_1, shown on the left. On the right are the output signal S_2 and noise N_2, after amplification G. The system has an intrinsic noise contribution T_R. For a direct detection device, $G=1$

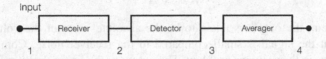

Fig. 4.4 This is needed for problem 12; it is also Fig. 4.7 of 'Tools'. The figure shows the components of a receiver

if a room-temperature load is connected to the receiver input. Suppose $F = 2$, what is T_R? Repeat for $F = 1.2$ and 1.5.

12*. The analysis in Section 4.2.1 of 'Tools' is applied to a square law detector. This analysis, taken from 'Tools', is given here. Modify this analysis, step for step, to derive the response of a *linear* detector. The relevant receiver block diagram is shown in Fig. 4.7 of 'Tools', shown here as figure 4.4. The relevant steps in this analysis for a square law detector are:

$$P_2 = v_2^2 = \sigma^2 = k\, T_{\text{sys}}\, G\, \Delta v ,$$

where Δv is the receiver bandwidth, G is the gain, and T_{sys} is the total noise from the input T_A and the receiver T_R. The output of the square law detector is v_3: $\langle v_3 \rangle = \langle v_2^2 \rangle$. After square-law detection we have

$$\langle v_3 \rangle = \langle v_2^2 \rangle = \sigma^2 = k T_{\text{sys}} G\, \Delta v .$$

The noise is the mean value and variance of $\langle v_3 \rangle$, for Gauss functions is:

$$\langle v_3^2 \rangle = \langle v_2^4 \rangle = 3 \langle v_2^2 \rangle$$

this is needed to determine $\langle \sigma_3^2 \rangle$. Then, from the definition of variance:

$$\sigma_3^2 = \langle v_3^2 \rangle - \langle v_3 \rangle^2$$

$\langle v_3^2 \rangle$ is the total noise power (= receiver plus input signal). Using the Nyquist sampling rate, the averaged output, v_4, is $(1/N)\Sigma v_3$ where $N = 2\Delta v\,\tau$. From v_4 and $\sigma_4^2 = \sigma_3^2/N$, the result is

$$\sigma_4 = k\Delta v\, G\, (T_A + T_R)/\sqrt{\Delta v\,\tau}$$

Use the calibration procedure to eliminate the term $kG\Delta v$. Apply this to a linear detector.

$$\frac{\Delta T}{T_{sys}} = \frac{1}{\sqrt{\Delta v\,\tau}}$$

In this system, the output is taken to be the absolute value of the voltage input. Assume that the signal is small compared to the receiver noise. Complete each calculation as in the previous problem. The output of the linear detector is

$$v_3 = \int |v_2|\exp(-v_2^2/2\sigma_2^2)\mathrm{d}x\ ,$$

while the noise depends on $\langle v_3 \rangle^2 = \langle v_2 \rangle^2 = \sigma^2$.
To obtain the final result, one must make use of the relation (fig. 4.5).

$$\Delta T_{RMS} = \frac{\sigma_4}{(\Delta \langle v_4 \rangle / \Delta T_s)}\ .$$

13. The Y factor is used to determine receiver noise. Given that T_L is $77\,\mathrm{K}$ and $T_H = 290\,\mathrm{K}$, show that the plot in Fig. 4.11, given here, correctly expresses the relation between T_{rx} and the Y factor.

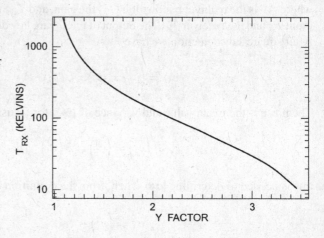

Fig. 4.5 This is Fig. 4.11 from 'Tools'. It is used in problem 13. This is a plot of receiver noise versus Y factor

14. Suppose a receiver accepts inputs from two frequencies, v_u and v_l. The response of the receiver is the same at these frequencies.
(a). If all gain and loss factors are equal, and the signal is present in both v_u and v_l, how does the value of T_R change?
(b). Suppose the signal is present in v_u only. Repeat part (a).
(c). Repeat (b) for the situation in which the response at v_u is twice that at v_l. What is the value of T_R?

15. Derive Eq. (4.57), which is:

$$\tau_m = \frac{1}{\sqrt{\Delta v \, \gamma_1}}.$$

from Eq. (4.56) of 'Tools':

$$\frac{\Delta T}{T_{sys}} = K\sqrt{\frac{1}{\Delta v \, \tau} + \left(\frac{\Delta G}{G}\right)^2}.$$

and the assumed fluctuations in receiver gain, G:

$$\left(\frac{\Delta G}{G}\right)^2 = \gamma_0 + \gamma_1 \tau,$$

16. To detect a source one samples a large region of the sky. The receiver is perfectly stable. If one has 10 samples at the position of the source, and 10^3 samples away from the source. If this involves spectral lines, one can fit a curve to the off-source data and subtract this from the on-source data. Justify the assertion the if the RMS noise of the on-source data is N, the noise in the difference of on-source and off-source is $N\sqrt{1 + 0.01}$.

17. What is the minimum receiver noise possible with a coherent receiver operating at 115 GHz? At 1000 GHz? At 10^{14} Hz.

18. Derive the results for Δv and τ in Table C.1 of 'Tools', given here as table 4.2. To carry out this evaluation, you need the definitions: a signal detectable if the mean output increment is greater than or equal to the dispersion of z; i.e. if $\langle \Delta z \rangle \geq \sigma_z$. Generally, the smoothing filter output power transfer function is only a few Hertz wide.

$$\frac{\Delta z}{\langle z \rangle} = \sqrt{\frac{2 \int_{-\infty}^{\infty} |G(v)|^2 \, dv \, \int_{-\infty}^{\infty} W(v) \, dv}{\left(\int_{-\infty}^{\infty} G(v) \, dv\right)^2 \, W(0)}}.$$

Table 4.2   Equivalent bandwidth of some filters and time smoothing (problem 18)

Reception filter	$G(v)$	Δv
Rectangular pass band	$\begin{cases} 1 \text{ for } v_0 - \frac{1}{2}\Delta < \lvert v \rvert < v_0 + \frac{1}{2}\Delta \\ 0 \text{ elsewhere} \end{cases}$	Δ
Single tuned circuit	$[1 + (\lvert v \rvert - v_0)^2/\Delta^2]^{-1}$	$2\pi\Delta$
Gaussian pass band	$\exp[-(\lvert v \rvert - v_0)^2/2\Delta^2]$	$2\sqrt{\pi}\,\Delta$
Smoothing filter	$W(v)$	τ
Running mean over time T	$(\pi T v)^{-2} \sin^2 \pi T v$	T
Single RC circuit	$[1 + (2\pi RC v)^2]^{-1}$	$2RC$
Rectangular pass band	$\begin{cases} 1 \text{ for } \lvert v \rvert < v_0 \\ 0 \text{ for } \lvert v \rvert > v_0 \end{cases}$	$\frac{1}{2}v_0$
Gaussian pass band	$\exp\left\{-v^2/2v_0^2\right\}$	$\sqrt{2\pi}\,v_0$

The first term depends only on the predetector bandwidth of the receiver, while the second term depends on the smoothing filter time constant. Thus defining bandwidth as:

$$\Delta v = \frac{1}{2} \frac{\left(\int\limits_{-\infty}^{\infty} G(v)\,dv \right)^2}{\int\limits_{-\infty}^{\infty} \lvert G(v) \rvert^2\,dv}$$

and the smoothing is

$$\tau = \frac{W(0)}{\int\limits_{-\infty}^{\infty} W(v)\,dv}$$

Note that $\Delta z \propto \Delta T$ and $\langle z \rangle \propto T_{\text{sys}}$, so we have:

$$\frac{\Delta T}{T_{\text{sys}}} = \frac{1}{\sqrt{\Delta v\,\tau}}$$

Chapter 5
Practical Receiver Systems

1. Determine the slope of the minimum receiver noise in Fig. 5.13 of 'Tools', given here as figure 5.1.

2. Coherent and incoherent receivers are fundamentally different. However one can determine the equivalent noise temperature of a coherent receiver T_n which corresponds to the NEP of a bolometer. This can be determined by using the relation

$$\text{NEP} = 2kT_n \sqrt{\Delta \nu} \,.$$

For $\Delta \nu = 50\,\text{GHz}$, determine T_n for $\text{NEP} = 10^{-16}\,\text{W}\,\text{Hz}^{-1/2}$. A bolometer receiver system can detect a 1 mK source in 60 s at the 3σ level. The bandwidth is 100 GHz. How long must one integrate to reach this RMS noise level with a coherent receiver with a noise temperature of 50 K, and bandwidth 2 GHz?

3. In the millimeter and sub-millimeter range, the Y factor (see, e.g. problem 13 of chapter 4) usually represents a double-sideband system response. See Fig. 4.11 of 'Tools', given as figure 4.5 in chapter 4. For spectral lines, one wants the single-sideband receiver noise temperature. If the sideband gains are equal, what is the relation of the Y factor for a single- and double-sideband system?

4. The definition of a *decibel*, db, is

$$\text{db} = 10 \log \left(\frac{P_{\text{output}}}{P_{\text{input}}} \right).$$

If a 30 db amplifier with a noise temperature of 4 K is followed by a mixer with a noise temperature of 100 K, what is the percentage contribution of the mixer to the noise temperature of the total if (see Eq. (5.17) in 'Tools', given below)

$$T_{\text{sys}} = T_{\text{stage}\,1} + T_{\text{stage}\,2}/\text{Gain}_{\text{stage}\,1}$$

© Springer International Publishing AG, part of Springer Nature 2018
T. L. Wilson, S. Hüttemeister, *Tools of Radio Astronomy – Problems
and Solutions*, Astronomy and Astrophysics Library,
https://doi.org/10.1007/978-3-319-90820-5_5

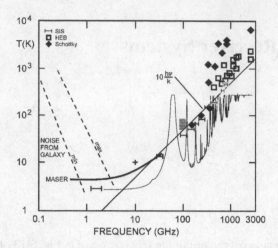

Fig. 5.1 This is Fig. 5.13 from 'Tools'. The receiver noise temperatures for coherent amplifier systems compared to the temperatures from different astronomical sources and the atmosphere. The atmospheric emission is based on a model of zenith emission for 0.4 mm of water vapor (plot from B. Nikolic (Cambridge Univ.) from the "AM" program of S. Paine (Center for Astrophys.)). This does not take into account the absorption corresponding to this emission. In the 1–26 GHz range, the horizontal bars represent the noise temperatures of HEMT amplifiers. The shaded region between 85 and 115.6 GHz is the receiver noise for the SEQUOIA array which is made up of monolithic millimeter integrated circuits (MMIC), at Five College Radio Astronomy Observatory. The meaning of the other symbols is given in the upper left of the diagram. For the SIS mixers, we have used the ALMA specifications. These are single sideband mixers covering the frequency range shown by the horizontal bars. The mixer noise temperatures given as double sideband (DSB) values were converted to single sideband (SSB) temperatures by increasing the receiver noise by a factor of 2. The ALMA mixer noise temperatures are SSB. The HEMT values are SSB

5. (a) In Fig. 5.5 of 'Tools', shown here as figure 5.2. In this figure, the upper sideband (USB) frequency is 115 GHz, and the lower sideband frequency is 107 GHz. What is the intermediate frequency? What is the Local Oscillator (LO) frequency?

(b) When observing with a double-sideband coherent receiver, an astronomical spectral line might enter from either upper or lower sideband. To distinguish between these two possibilities, one uses the following procedure. To decide whether the line is actually in the upper or lower sideband, the observer increases the local oscillator frequency by 100 kHz. The signal moves to *lower* frequency. Is the spectral line from the upper or lower sideband?

6. The same situation as in Problem 5, but after the first mixer is a second mixer with an LO frequency which is *higher* than the intermediate frequency of the first mixer. The spectral line is known to be in the upper sideband. To eliminate unwanted spectral lines, someone tells you to move the LO higher frequencies in steps of 100 kHz, and at the same time, move the next oscillator in the LO chain, LO2, to lower frequencies by the same step. After repeating this procedure for 10 steps of 100 kHz, the result is added. Will this procedure eliminate spectral lines in the lower

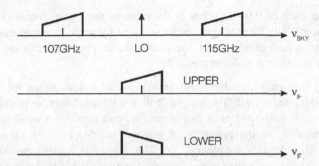

Fig. 5.2 This is Figure 5, Chapter 5 in 'Tools'. Here it refers to problem 5. This is a sketch of the frequencies shifted from the sky frequency (*top*) to the output (*lower*) of a double sideband mixer. In this example, the input is at the sky frequencies for the Upper Side Band (USB) of 115 GHz, and Lower Side Band (LSB) of 107 GHz while the output frequency is 4 GHz. The slanted boxes represent the passbands; the direction of the slant in the boxes indicate the upper (*higher*) and lower (*lower*) edge of the bandpass in frequency

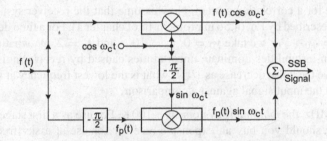

Fig. 5.3 This is Figure 5.6 from 'Tools'. This refers to problem 7. It is a sketch of the single sideband mixer (SSB). The input signal, $f(t)$, is divided into two equal parts. There are two identical mixers located in an *upper* and *lower* branch of the sketch. The monochromatic LO frequency from a central source, ω_c, is shifted in phase by $\pi/2$ from the input to the output of the mixer in the *lower part* of the sketch. In the *lower* branch, the phase of the input signal is also shifted by $\pi/2$. After mixing the signals are added to produce the single sideband output. For further details, see problem 7

sideband? If the unwanted lower sideband spectral line has a width of 100 kHz, by how much is this line reduced in intensity?

7. In Fig. 5.6 of 'Tools', shown here as figure 5.3, we give the schematic of a *single-sideband mixer*. In such a system, the image and signal bands are separated in the output if the input is $f(t) = \cos \omega_s t$. Use an analysis for this input signal to show that such a mixer is feasible. Repeat for $f(t) = \sin \omega_s t$.

8. The input power of a receiver can be 10^{-16} W, while the power at the output of a receiver must be about a milli Watt. What must be the power amplification of such a receiver? Express this in decibels. Suppose the gain stability of this receiver is 10^{-3} over 30 s. What is the change in the output power? Suppose that the system noise is 100 K and the bandwidth is 1 GHz. This is used to measure a source with a peak temperature of 0.01 K. What is the ratio of the signal intensity to that of gain fluctuations? The fluctuations can be reduced by periodic comparisons with a reference source; how often should one switch the receiver between the signal and a reference to stabilize the output power?

9. Laboratory measurements frequently make use of a data-taking method which involves a modulated signal. The output is then measured synchronously with the modulation rate in both frequency and phase. We can measure a weak input signal, $S = T(\text{signal})e^{-\tau}$, in the presence of noise, $T(\text{cable})(1-e^{-\tau})$, by modulating the signal with a known frequency, f_1. The output is superimposed on noise background. What is the noise in the switched output? What is the signal-to-noise ratio? How will the signal-to-noise ratio change with time if only random noise is present?

10. This makes use of the equations in problem 15 of chapter 4. If the bandwidth of a receiver is 500 MHz, how long must one integrate to reach an RMS noise which is 0.1% of the system noise with a total power system? Repeat for a Dicke switched system, and for a correlation system. Now assume that the receiver system has an instability described by Eq. (4.56 in problem 15 of chapter 4). For a time dependence $(\Delta G/G)^2 = \gamma_0 + \gamma_1\tau$ we take $\gamma_0 = 0$, $\gamma_1 = 10^{-2}$ and $K = 2$. On what time scale will the gain instabilities dominate uncertainties caused by receiver noise? If one wants to have the noise decrease as $1/\sqrt{t}$, what is the lowest frequency at which one must switch the input signal against a comparison?

11. At 234 MHz, the *minimum* sky noise is ~100 K. For use as a first stage amplifier at 234 MHz should you buy an expensive receiver for use at a sky frequency of 234 MHz which has a noise temperature of 10 K, if a similar receiver has a noise temperature of 50 K but costs 10% of the price of the lower-noise receiver? Explain your decision by considering observational facts.

12. An all-sky continuum survey covering 41,252 square degrees, is carried out with a 40' (=0.44 square degrees) beam at 234 MHz. Three spatial samples are taken for each beamwidth. These samples are used to image the sky at 234 MHz.
(a) Compare the sampling procedure to the Nyquist sampling rate using the example of the sampling of sine or cosine waves. What is the total number of samples?
(b) Next, assume that the sky noise dominates the receiver noise. If the bandwidth B is 10 MHz and the integration time is 10 s per position, what is the RMS noise as a fraction of T_{source}, the sky noise? How many data points are needed to completely characterize the resulting map? If one needs 20 s of time for measuring each position, how long will this survey require?
(c) Repeat this estimate for a survey at 5000 MHz carried out with a 3' beam, for a receiver with noise temperature 50 K, 500 MHz bandwidth, 10 s integration per point. Note that the sky background contributes only a small amount of the receiver noise at 5 GHz. How much observing time is needed for this survey?

Chapter 6
Fundamentals of Antenna Theory

1. Complete the mathematical details of summing the expression in Eq. (6.49) of 'Tools', which is given here:

$$\hat{S} = \Sigma_{n=0}^{N} e^{i\,kn\,D\,\sin(\phi)}$$

First, show that

$$\hat{S} = \Sigma_{n=0}^{N} q = \frac{1 - q^{n+1}}{1 - q}$$

(hint: multiply by q to obtain another relation, then subtract from the relation above). With $q = e^{i\,k\,D\,\sin(\phi)}$, show that we obtain Eq. (6.50), which is:

$$\hat{S} = e^{i\,k\,D\,\sin(\phi)} \cdot e^{-i\,(N-1)\,k\,D/2\,\sin(\phi)} \cdot \left[\frac{\sin\frac{kND}{2}\sin(\phi)}{\sin\frac{kD}{2}\sin(\phi)} \right]$$

From this, one can obtain the power pattern:

$$\hat{S}^2 = \left[\frac{\sin[k\,(n+1)D/2]\sin(\phi)}{\sin[k\,D/2]\sin(\phi)} \right]^2$$

Use limits to show that the square of the x component of Eq. (6.64) in 'Tools' can be obtained from the expression above. This component is:

$$P_n(l) = \left[\frac{\sin(\pi l\,L_x/\lambda)}{\pi l\,L_x/\lambda} \right]^2$$

© Springer International Publishing AG, part of Springer Nature 2018
T. L. Wilson, S. Hüttemeister, *Tools of Radio Astronomy – Problems
and Solutions*, Astronomy and Astrophysics Library,
https://doi.org/10.1007/978-3-319-90820-5_6

2. You read that there are antennas without sidelobes. That is, *all* of the energy is contained in the main lobe. Should you believe the report? Comment using qualitative arguments, but *not* detailed calculations.

3. If the size of the pupil of the human eye, D is 0.5 cm, what are the number of wavelengths across this aperture for light of $\lambda = 500$ nm? Compare this to the number of wavelengths across the aperture of a 100 m radio telescope for a wavelength of 2 m, 1 cm. Repeat for the ALMA radio telescope, with a diameter of 12 m, for $\lambda = 1$ cm, 3 mm, 0.3 mm. Discuss the implications of these results.

4. Derive the increase in the radiated power for an array of N dipoles, for the case of phases set to zero in Eq. (6.51 in 'Tools'). This is:

$$|\hat{S}|^2 = \left[\frac{\sin\left(\frac{kND}{2}\sin(\phi)\right)}{\sin\left(\frac{kD}{2}\sin(\phi)\right)} \right]^2$$

Compare this to the maximum power radiated in a given direction by a single Hertz dipole, expressed in these terms as $|\hat{S}|^2 = [\sin\phi]^2$.

5. The full width half power (FWHP) angular size, θ, in radians, of the main beam of a diffraction pattern from an aperture of diameter D is $\theta \approx 1.02\lambda/D$.
(a) Determine the value of θ, in arc min, for the human eye, where $D = 0.3$ cm, at $\lambda = 5 \times 10^{-5}$ cm.
(b) Repeat for a filled aperture radio telescope, with $D = 100$ m, at $\lambda = 2$ cm, and for the very large array interferometer (JVLA), $D = 27$ km, at $\lambda = 2$ cm.
(c) Show that when λ has the units of millimeters, and D the units of kilometers and θ the units of arc seconds, then $\theta = 0.2\lambda/D$. Is this consistent with Eq. (6.78), which is $\theta = 58.4 \times \frac{\lambda}{D}$?

6. Hertz used $\lambda \approx 26$ cm for the shortest wavelength in his experiments.
(a) If Hertz employed a parabolic reflector of diameter $D \approx 2$ m, what was the FWHP beam size? (See Problem 3 of this chapter.)
(b) If the $\Delta l \approx 26$ cm, what was the radiation resistance, from Eq. (6.43)? Eq. (6.43) is:

$$R_S = \frac{c}{6}\left(\frac{\Delta l}{\lambda}\right)^2$$

(c) Hertz's transmitter was a spark gap. Suppose the current in the spark was 0.5 A. What was the average radiated power?

7. Over the whole world, there have been (on average) 100 radio telescopes of (average) diameter 25 m operating since 1960. Assume that the power received by each is 10^{-15} W over this period of time. What amount of energy has been received in this period of time? Compare this to the energy released by an ash (taken to be 1 g) from a cigarette falling a distance of 2 cm in the earth's gravity.

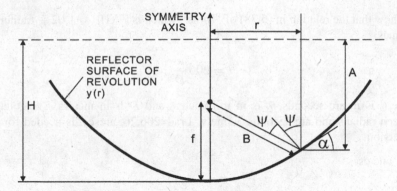

Fig. 6.1 This is Fig. 6.11 from 'Tools'. This is a sketch of a parabola showing angles used in problem 8

8. Refer to figure 6.1 which is Fig. 6.11 of 'Tools'. The surface is described by $y(x) = (1/4f)x^2$.

(a) Find a general expression for the path from the pupil plane (dashed line) to the focus, f.

(b) If an on-axis plane wave is in the pupil plane, show that for a paraboloid, there is a single focus.

(c) Is there such a relation for a circle, $y(x) = \sqrt{R_0^2 - x^2}$?

9. Show that the two dimensional expression for the **E** field Eq. (6.56) in 'Tools', given here as:

$$dE_y(\phi) = -i\, J_0\, g(x')\, \frac{1}{r} e^{-i(\omega t - kr)}\, dx'\ .$$

and one of the factors in the three dimensional diffraction Eq. (6.57), given here as:

$$dE_y = -\frac{i}{2}\lambda J_0\, g(\mathbf{x}')\frac{F_e(\mathbf{n})}{|\mathbf{x} - \mathbf{x}'|} e^{-i(\omega t - k|\mathbf{x} - \mathbf{x}'|)}\, \frac{dx'}{\lambda}\frac{dy'}{\lambda}\ .$$

are related by equating $(|x - x'|) = \frac{1}{r}$ and identifying J_0 as $I\Delta l/2\lambda$, the current density.

10. If two dipoles spaced by $\lambda/4$ are connected to a coherent input, what is the far field radiation pattern if the phases of the dipoles differ by $\lambda/4$? Simplify by using a one dimensional geometry.

11. Suppose you have a single dipole at $\lambda/4$ in front of a perfectly conducting plate. Determine the far field radiation pattern. Compare this to the result of problem 10.

12. Show that the relation in (6.78) of 'Tools', which is FWHP $= 1.02 \frac{\lambda}{D}$ radians becomes

$$\theta = 206 \frac{\lambda}{D}$$

where θ is in arc seconds, λ is in millimeters and D is in meters. The relation between radians and arc seconds, namely, 1 rad=206,265 arcsec, is needed for the conversion.

13. If one uses

$$\frac{I \, \Delta l}{2\lambda} = \frac{e \, \dot{V}^2}{c^2}$$

show that the expression in Eq. (6.41), which is $P_S = \frac{c}{3} \left(\frac{I \Delta l}{2\lambda} \right)$ becomes:

$$P(t) = \frac{2}{3} \frac{e^2 \dot{V}}{c^2}$$

this is the radiation from an accelerated electron of charge e and velocity V, from classical electromagnetic theory. The relation for the radiation from the bound states m and n, where ν_{mn} is the line frequency and $|\mu_{mn}|$ is the *dipole moment*, from quantum theory is:

$$\langle P \rangle = \frac{64\pi^4}{3\,c^3} \nu_{mmn}^4 \left(|\mu_{mn}| \right)^2$$

Determine the relation of the quantum mechanical to the classical result.

14. In Eq. (6.40), which is:

$$|\langle S \rangle| = \frac{c}{4\pi} \mid \text{Re} \, (\mathbf{E} \times \mathbf{H}^*) \mid = \frac{c}{4\pi} \left(\frac{I \Delta l}{2\lambda} \right)^2 \frac{\sin^2 \vartheta}{r^2}$$

use the more realistic variation of current in the dipole:

$$I(z) = I_0 \times \left(1 - \frac{|z|}{\Delta l / 2} \right)$$

to determine the power radiated and the radiation resistance (Eq. (6.43)). Show that R_S is

$$R_S = \frac{c}{24} \left(\frac{\Delta l}{\lambda} \right)$$

Chapter 7
Practical Aspects of Filled Aperture Antennas

1. (a) Use Eq. (7.3) in 'Tools', which is: $\Omega_A = \int_0^{2\pi} \int_0^{\pi} P_n(\vartheta, \varphi)\, d\Omega$,
Eq. (7.5) in 'Tools', which is: $\Omega_{MB}/\Omega_A = \eta_B$
Eq. (7.20) in 'Tools', which is: $\mathcal{P}_\nu = \frac{1}{2} A_e \iint B_\nu(\vartheta, \varphi) P_n(\vartheta, \varphi)\, d\Omega$
Eq. (7.21) in 'Tools', which is: $\mathcal{P}_\nu = k\, T_A$
and Eq. (7.23) of 'Tools', which is: $T_A(\vartheta_0, \varphi_0) = \dfrac{\int T_B(\vartheta,\varphi) P_n(\vartheta - \vartheta_0, \varphi - \varphi_0) \sin\vartheta\, d\vartheta\, d\varphi}{\int P_n(\vartheta,\varphi)\, d\Omega}$
Eq. (7.23) is the convolution of the telescope power pattern, P_n with the actual source distribution, T_B. The result is the observed source distribution.
Use these to show that $T_A = \eta_B T_B$, where T_B is the observed brightness temperature. to show that for a source with an angular size $\ll$ the telescope beam, $T_A = S_\nu A_e/2k$. Use these relations to show that $T_A = \eta_B T_B$, where T_B is the observed brightness temperature.
(b) Suppose that a Gaussian-shaped source has an actual angular size θ_s and actual peak temperature T_0. This source is measured with a Gaussian-shaped telescope beam size θ_B. The resulting peak temperature is T_B. The flux density, S_ν, integrated over the entire source, must be a fixed quantity, no matter what the size of the telescope beam. Use this argument to obtain a relation between temperature integrated over the telescope beam,

$$T_B = T_0 \left(\frac{\theta_s^2}{\theta_B^2 + \theta_s^2} \right)$$

Show that when the source is small compared to the beam, the main beam brightness temperature $T_B = T_0(\theta_s/\theta_B)^2$, and further the antenna temperature $T_A = \eta_B T_0 (\theta_s/\theta_B)^2$.

2. Suppose that a source has $T_0 = 600\,\mathrm{K}$, $\theta_0 = 40''$, $\theta_B = 8'$ and $\eta_B = 0.6$. What is T_A? (Use the result of Problem 1(b) of this chapter.)

© Springer International Publishing AG, part of Springer Nature 2018
T. L. Wilson, S. Hüttemeister, *Tools of Radio Astronomy – Problems and Solutions*, Astronomy and Astrophysics Library,
https://doi.org/10.1007/978-3-319-90820-5_7

3. Suppose your television needs $1\,\mu W$ of power at the input for good reception. The transmitter radiates $100\,kW$ in all azimuthal directions, and within an angle $\pm 10°$ about the horizontal direction, and is at $100\,m$ elevation. Ignore reflections and assume that the earth is perfectly flat. Calculate the effective area, A_e, that your TV antenna must have if you live $30\,km$ from the transmitter.

4. Suppose that your antenna has a normalized peak power, P, with the following values: $P = 1$ for $\theta < 1°$, $P = 0.1$ for $1° < \theta < 10°$, and $P = 0$ for $\theta > 10°$. What is Ω_A, from Eq. (7.3), given in problem 1 of this chapter? What is Ω_{MB} and η_B.

5. A scientist claims that for a very special antenna the brightness temperature of a compact source can exceed the antenna temperature. Do you believe this?

6. You are told that there is a special procedure which allows the *measured* Gaussian source size (*not* the deconvolved size) to be smaller than the Gaussian telescope beam. This can occur (so the claim goes) if the source is very intense. Do you believe this? This is the 'diffraction limit'. [In recent years, Nobel Prizes in chemistry have been given for violating this limit, but only with the aid of non-equilibrium excitation.]

7. The Gaussian function considered in chap. 4 was:

$$y(x) = A \, \exp\left(-\frac{x^2}{2\sigma^2}\right),$$

where A is a normalization constant. For radio astronomical applications, one usually takes the form of this function as

$$y(x) = A \, \exp\left(-\frac{4\ln2(x - x_0)^2}{\theta_{1/2}^2}\right).$$

Relate the parameters σ and $\theta_{1/2}$. The quantity $\theta_{1/2}$ is the FWHP, full width to half power. In the literature, the "width" of a Gaussian function is usually the FWHP.

8. The ground screen for the Arecibo telescope has a width of $15\,m$, and is mounted around the edge of the $305\,m$ diameter radio telescope. Assume you could direct the entire ground screen so that the power is collected at a single location.
(a) What is the geometric area of this ground screen? Take the antenna as a ring, with an inner radius of $305\,m$, the outer radius being $315\,m$.
(b) Calculate the far-field antenna pattern. What are the location and intensity in the first sidelobe, relative to the main lobe?
(c) Calculate the conversion factor, from Jy to K, for the antenna temperature if the antenna efficiency is 0.6.

9. Single telescope pointing is checked by scanning through the center positions of known sources by a few beamwidths in orthogonal directions. The positional error, $\Delta\theta$, caused by random noise, as measured with a beam of FWHP size θ_0 and signal-to-noise ratio of (S/N) is $\theta_0/(S/N)$. Neglect all systematic errors. What would have to be the (S/N) to determine a source position to 1/50 of the FWHP beamwidth of the telescope? Is there a contradiction between the angular resolution of a telescope, $\theta \sim \lambda/D$, and the positional accuracy?

10. Figure 7.6d in 'Tools', given in figure 7.1, represents off-axis reflectors such as the South Pole Telescope (SPT). This has a beam efficiency of 0.95 at a wavelength of 3 mm. Assume that $p = 0$, $K = 0$, as given in Table 6.1 of 'Tools' (i.e. no taper of the power pattern; a part of Table 6.1 is given here as Table 7.1). The relevant equation is:

$$P_n(u) = \left[\frac{2^{p+1}\, p!\, J_{p+1}(\pi u D/\lambda)}{(\pi u D/\lambda)^{p+1}} \right]^2$$

where J_{p+1} is a Bessel function of order $p + 1$, D is the dish diameter an λ is the wavelength. The parameter u is a parameter to specify the angular offset from the telescope axis.
From Figs. 7.6d and 7.7 of 'Tools' (our figure 7.1), what must be the surface accuracy?

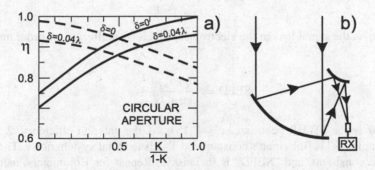

Fig. 7.1 A combination of Fig. 7.6d in (**a**) and 7.7 in (**b**), taken from 'Tools'. In (in (**a**)), the Aperture efficiency η_A (——) and beam efficiency η_{MB} (– – –) for different values of K in Table 6.1, reproduced in Table 7.1, here. The values for both an ideal reflector ($\delta = 0$) and one that introduces random phase errors of $\delta = 0.04\,\lambda$ are given. This diagram applies to an antenna with no blockage. Such antennas are the Green Bank Telescope, GBT and the South Pole Telescope, SPT [(**a**) after Nash (1964 *IEEE Trans. on Antenna Propag.* 12, 918)]

Table 7.1 Normalized power pattern characteristics produced by aperture illumination (problem 10)

p	K	FWHP (rad)	BWFN (rad)	Relative gain	First side lobe (dB)
0		1.02	2.44	1.00	−17.6
1		1.27	3.26	0.75	−24.6

(a) What must be the antenna efficiency from the figure?

(b) At one time, a similar telescope was used for satellite tests at 28 GHz. The satellite is a point source in the beam of this telescope, so η_A should be optimized for a point source. Now what are the values of antenna and beam efficiency? What is the beam size?

11. Combine Eqs. 7.5, which is $\eta_B = \frac{\Omega_{MB}}{\Omega_A}$, Eq. (7.9), which is, $A_e = \eta_A A_g$ and Eq. (7.11), which is $A_e \Omega_A = \lambda^2$ together with $\theta_g = \theta_{geom} = \frac{\lambda}{D}$ and $A_{geom} = \frac{\pi}{4} D^2$ to obtain the relation

$$\eta_B = 1.133 \times \eta_A \frac{\pi}{4} \left[\frac{\theta_B}{\theta_{geom}} \right]^2$$

Use this result to determine the value of θ_B in terms of θ_{geom} for the highest and lowest values of η_A and for the value at the point when $\eta_B = \eta_A$.

12. Use Eq. (6.52), which is $\theta = \lambda/D$, Eq. (7.9), which is given in problem 11, and the definition of SEFD (Eq. (7.27)), which is

$$SEFD = \frac{2 \eta' k T_{sys}}{A_{eff}}$$

where η' is the signal loss in the electronics, which is small. η' is of order unity, to obtain:

$$SEFD = \frac{2\eta' k T_{sys} \theta^2}{\pi \eta_a \lambda^2}$$

where θ is the FWHP beamsize, $\eta' \approx 1$, η_a is the antenna efficiency, λ is the wavelength, "k" is Boltzman's constant and T_{sys} is the total system noise. The units must be consistent, and "SEFD" is in Janskys. Repeat for Bolometers, using the relation of NEP to $T_{sys} = T_{BG}$ as in Eq. (5.13), that is:

$$NEP = 2 \varepsilon k T_{BG} \sqrt{\Delta \nu}$$

13. Use Eq. (6.40), namely $P(\theta) = P_0 \sin^2 \theta$ to determine the normalized power pattern of the Hertz dipole. Use Eq. (7.2), which is:

$$G(\vartheta, \varphi) = \frac{4\pi P (\vartheta, \varphi)}{\iint P (\vartheta, \varphi) \, d\Omega}$$

where the integral is over 4π.

Use Eqs. 7.3, 7.5 and 7.11, given in problems 1 and 11 of this chapter, to obtain Ω_A, Ω_{MB}, η_B and A_e.

14. Use Eqs. 7.9–7.11, given previously, to obtain the effective area of the Hertz dipole using a linear variation of current along the dipole,

$$I(z) = I_0 \times \left(1 - \frac{|z|}{\Delta l/2} \right)$$

(this is Eq. (6.86) in 'Tools' and problem 14 of chapter 6). Show that the effective area in this case is

$$A_e = \frac{3}{8\pi} \lambda^2$$

15. Calculate the *Rayleigh distance*, k, defined as $k = 2D^2/\lambda$, for an antenna of diameter $D = 100\,\mathrm{m}$ and a wavelength $\lambda = 3\,\mathrm{cm}$. This is also referred to as the *Frauenhofer Distance*. It is one of the criteria that defines the far field, or wave zone of an antenna.

16. For a 305 m diameter radio telescope with $\eta_A=0.5$, what is the ratio of antenna temperature to flux density for a point source? for an antenna of diameter $D = 100\,\mathrm{m}$ and a wavelength $\lambda = 3\,\mathrm{cm}$.

Chapter 8
Single Dish Observational Methods

1. Investigate the effect of the earth's atmosphere on radio observations by using a single layer atmosphere, using Eq. (1.37), which is:

$$T_b(s) = T_b(0) + T(1 - e^{-\tau_\nu(s)})$$

Suppose we know that the atmospheric optical depth, τ, is 0.1, and the temperature is 250 K.
(a) What is the excess noise from the atmosphere, and what is the reduction in the intensity of a celestial source?
(b) Repeat for $\tau = 0.5, 0.7, 1.0, 1.5$.
(c) If τ is related to the optical depth in the zenith by $\tau = \tau_z/\sin(\text{elv})$, determine the increase in τ between 30° and 20° elevation. (Elevation is measured relative to the horizon.)
(d) Repeat this calculation for the increase between 20° and 19°, then 20° and 15°.
(e) For spectral line measurements, one is interested in a comparison of the responses of the receiver system over a (relatively) small frequency interval. Consider the measurement of a 10 mK spectral line through an atmosphere with $\tau = 0.2$, if the receiver noise is 100 K. Repeat this calculation for a receiver noise of 20 K.

2. A standard method to determine atmospheric τ values employs a receiver to determine the emission of the earth's atmosphere at 225 GHz. Suppose this emission is found to be 15 K at elevation 90°, 18 K at 60°, 30 K at 30°, and 42 K at 20°. If the temperature of the atmosphere is 250 K, what is the zenith τ? Is the curve in Fig. 8.1 consistent with ratios of zenith τ to that at 225 GHz are 3.4 (at 340 GHz), 6.7 (at 410 GHz), 9.9 (at 460 GHz) and 19.0 (at 490 GHz).

3. Suppose you are observing at 1 cm wavelength with a filled aperture telescope. When pointed toward cold sky, in the zenith, your system noise temperature is twice

© Springer International Publishing AG, part of Springer Nature 2018
T. L. Wilson, S. Hüttemeister, *Tools of Radio Astronomy – Problems
and Solutions*, Astronomy and Astrophysics Library,
https://doi.org/10.1007/978-3-319-90820-5_8

what you expect. Normally the receiver noise temperature is 70 K and system noise
temperature is 100 K. Your partner notices that the radio telescope is filled with wet
snow. Assuming that the snow has a temperature of 260 K, and is a perfect absorber
at 1 cm, how much of the telescope surface is covered with snow?

4. A group observe sources at 1.3 cm at elevations between 8° and 11°. If the
zenith optical depth is $\tau_z = 0.1$, use an assumed dependence of $\tau=\tau_z/\sin(\text{elv})$ to
determine τ at the lowest and highest elevations. These astronomers see at most a
30% change in τ over this range of elevations. Is this reasonable? If the *receiver*
noise is 40 K, what is the *system* noise, including the atmospheric contribution, for a
200 K atmosphere, at these elevations? The observations are mostly of spectral lines;
how much is the attenuation? The temperature scale is calibrated using a nearby
source with peak main beam brightness temperature 16 K. What is the RMS error
for each continuum data point, from noise only, if the bandwidth used is 40 MHz
and the integration time is 1 s?

5. Use the Rayleigh–Jeans approximation to calculate the numerical relation
between flux density, S_ν and brightness temperature, T_B, if the source and beam
have Gaussian shapes. S_ν must be in units of Janskys ($= 10^{-26}\,\text{W}\,\text{m}^{-2}\,\text{Hz}^{-1}$),
wavelength must be in cm, and the observed angle θ_0 in arc min.

6. For a Gaussian-shaped source of actual angular size θ_{source} and observed
size θ_{observed}, find the relation between the apparent or main beam brightness
temperature, T_{MB}, and the actual brightness temperature, T_B. (Use the fact that
the flux density of a discrete source must not depend on the telescope.) Show that
$T_B > T_{MB}$. Show that the observed or apparent, actual and telescope beam sizes,
θ_{observed}, θ_{source} and θ_{beam}, are related by $\theta_{\text{observed}}^2 = \theta_{\text{actual}}^2 + \theta_{\text{beam}}^2$.

7. An outburst of an H_2O maser (at 22.235 GHz) in the Orion region (distance from
the Sun 500 pc) gave a peak flux density of 10^6 Jy over a 1 MHz band. If this maser
radiation were measured with the 100 m telescope, which has a collecting area of
$7800\,\text{m}^2$, and antenna efficiency 0.4, what is the peak power? If the safety level for
microwave radiation for humans is $10\,\text{mW}\,\text{cm}^{-2}$, at what distance would the Orion
maser be a threat for humans?

8. Use the Rayleigh–Jeans relation to calculate the flux density of the Sun at 30 GHz
if the disk has a diameter of 30′ at a uniform surface temperature 5800 K? Suppose
we had a 40 m radio telescope with effective collecting area $1000\,\text{m}^2$. What is the
value of T_{MB}? If $\eta_A = 0.5$ and $\eta_{MB} = 0.65$, what is T_A?

9. Use Eq. (8.19) (this equation is given in the solution of problem 5 of this chapter),
to determine the peak main beam brightness temperature of the planetary nebula
NGC7027 at 1.3 cm with the 100 m telescope ($S(\text{Jy}) = 5.4\,\text{Jy}$, $\theta_0 = 43''$).
(a) If the actual (assumed uniform) source size is $\theta_s = 10''$, use Eq. (8.21), which is:
$T_s = T_{MB} \times \frac{\theta_s^2+\theta_b^2}{\theta_s^2}$ to determine the actual source brightness temperature T_s. Then
use Eq. (1.37), as given in the problem 1 statement of this chapter, with $T_0 = T_B(0)$
and $T = T_\nu$ with $T_0 = 0$, and $T_\nu = 14,000$ K to determine the peak optical depth
of this region at 1.3 cm.

10. A celestial source has a flux density of 1 Jy at 100 MHz. If the angular size is 10″, and source and telescope beams are Gaussians, estimate the source brightness temperature in the Rayleigh–Jeans limit. Repeat this for an observing frequency of 1 GHz.

11. The planet Venus is observed at the distance of closest approach, a distance of 0.277 AU. The radius of Venus is 6100 km. What is the full angular width of Venus in arc seconds? Suppose the measured brightness temperature of Venus at 3.5 cm wavelength in a telescope beam of 8.7′ is 8.5 K. What is the actual surface brightness temperature of Venus?

12. In the sub-millimeter range, sky noise dominates, but one wants to have the most sensitive receivers possible. Is this a contradiction? If not, why not?

13. The APEX submillimeter telescope on the ALMA site has a diameter of 12 m, an estimated beam efficiency of 0.5 at a wavelength of 350 μm. At 350 μm the atmospheric transmission is 5%.
(a) Show that this is equivalent to a τ of 3.
(b) What is the sky noise for this situation if the physical temperature of the sky is 200 K?
(c) If the *receiver* noise is 50 K, what is the total *system* noise?
(d) Suppose you plan to measure a small diameter source with a flux density of 0.1 Jy. After what length of time will you have a signal-to-noise ratio of unity if the receiver bandwidth is 2 GHz?

14. Use the following expression for the redshift, z, which is based on the special theory of relativity. This can be used to relate the speed of expansion v to the z value $1 + z = \sqrt{\frac{c+V}{c-V}}$. Find V/c for z values of 2.28, 5, 1000.

15. Spectral line observations are carried out using position switching, that is the "on–off" observing mode. Thus effects of ground radiation should cancel in the difference spectrum. However, there is usually a residual instrumental baseline found in the case of centimeter wavelength observations. The amplitude of this residual instrumental baseline is found (with the 100 m telescope) to be $\sim 10^{-3}$ of the continuum intensity of the source being observed. This effect is caused by the correlation of signal voltage E_i, with that reflected by the primary feed horn, E_r. How much *power flux*, E_r^2, (in W m^{-2}) relative to E_i, is reflected from the feed?

16. A search for dense molecular gas in the Orion cloud shows the presence of 125 sources, each with a FWHP of 1′. The region searched is 15′ by 120′. If the beam size is 20″, what is the mean number of sources per angular area? Now use Poisson statistics $P = e^{-m} m^n / n!$ where n is the number of expected sources, and m is the mean, to find the probability of finding a dense clump of gas in this region if one uses a 20″ beam. What is the chance of finding two such sources?

17. (a) In an extragalactic survey, the average number of sources per beam is 0.04. Use Poisson statistics to find the chance of finding 2 or 3 sources in the same beam?
(b) Use these results to estimate the number of beam areas per source needed to insure that source confusion is a small effect.

18. Derive the result in Eq. (8.62), which is:

$$\sigma^2 = \left(\frac{q^{3-\gamma}}{3-\gamma}\right)^{\frac{1}{\gamma-1}} (k\,\Omega_e)^{\frac{1}{\gamma-1}}$$

showing all steps.

19. (a) For radio telescopes, the one dimensional power pattern is $y(x) = A\exp\left(-\frac{4\ln 2x^2}{\theta_{1/2}^2}\right)$ Use this expression to evaluate Eq. (8.59), which is:

$$\Omega_e = \int [f(\theta,\phi)]^{\gamma-1}\,d\Omega$$

(b) Calculate $k = \gamma\,N_c\,S_c^{-\gamma}$ for $\gamma = 2.5$, $q = 5$, $\Omega = 80 \times 100$ arcsec, $N_c = 10^5$ per steradian, and $S_c = 10^{-28}$ W m^{-2} Hz^{-1}=10^{-2} Jy.

20. The approach used by Mills and Slee (1957 Austral. J. Phys., 10, 162) illustrates the relation of the observed flux densities to number of sources for the Euclidean model of the universe. This approach is derived next. Start with the relation of flux density S to intensity emitted by a source at a distance R, P_0. Then we have $S = \frac{P_0}{4\pi R^2}$.

Show that the total number of sources in a sphere of radius R is: $n = \frac{4\pi}{3}R^3 n_0$ Show that this is equivalent to $n = k'S^{-1.5}$ Show that the differential number of sources is $dn = k'S^{-2.5}\,dS$. Evaluate k'.

21. This is a derivation that gives a result comparable to Eq. (8.62), which is given in problem 18. Here we use an approach that expresses the dispersion σ^2 in terms of limiting flux density rather than instrument deflection, D_c. This is sometimes referred to as the "Classical Approach to Confusion", as given in the 4th edition of "Tools of Radio Astronomy". Let $p(S)$ be the probability density for the number of sources per steradian with a flux density between S and $S + dS$; that is, on average we will observe $\bar{\nu} = \Omega\,p(S)\,dS$ sources with flux densities in the interval $(S, S+dS)$ per beam Ω. The average number of sources per beam is $\bar{n}$. If the sources are distributed according to a Poisson distribution (see, e.g. "An Introduction to Error Analysis, second edition", 1997, J. R. Taylor, University Science Books), the probability of n sources in a beam is $f(n) = \frac{\bar{\nu}^n}{n!}e^{-\bar{\nu}}$ Show that the second moment μ_2, of this distribution is $\mu_2 = \sum_{\nu=0}^{\infty} \nu^2 f(\nu) = \bar{\nu}(\bar{\nu}+1)$, so that the dispersion becomes $\sigma_\nu^2 = \bar{\nu}$. when applied to flux densities, show that this is $\sigma_s = S^2\,\bar{\nu}$.

(b) For simplicity, take the shape of the beam to be a "pill-box" with perpendicular walls, so that the total output of the antenna is the unweighted sum of the flux densities of all sources within the beam $S = \sum_k S_k$. Show that the average $\bar{S}$ caused by sources $(S, S + dS)$ will be $\bar{S} = S\bar{\nu}$. If the signal is caused by sources with different flux densities, the dispersions add quadratically. Making use of $\bar{v} = \Omega \, p(S) \, dS$ and $\sigma_S^2 = S^2 \bar{v}$, sources with flux densities between S_c and S_L and a number density of $p(S)$ then result in a dispersion for the total flux of $\sigma_{S_c}^2 = \Omega \int_{S_L}^{S_c} S^2 \, p(S) \, dS$. from Rice (1954 in *Selected Papers on Noise and Stochastical Processes*, N. Wax ed., Dover, New York).Then sources with a flux density cut-off S_c can be measured with a signal to noise ratio, q, of $q = \frac{S_c}{\sigma_{S_c}}$, where S_L is a lower limit to the actual flux density, whereas S_c is a cutoff

$$q^2 = \frac{S_c^2}{\Omega} \frac{1}{\int_{S_L}^{S_c} S^2 \, p(S) \, dS}$$ The distribution of faint point sources is $p(S) = n \, N_c \, S_c^n \, S^{-n-1}$

for a wide range of S with $n = 1.5$. Show that $q^2 = \left(0.333 \frac{1}{\Omega N_c} \frac{1}{1 - (S_L/S_c)^{0.5}} \right)$

(c) For the case $S_L = 0$, show that $\lim_{S_L \to 0} q^2 = 0.333 \frac{1}{\Omega N_c}$, Suppose $S_L = 0.666 \, S_c = 0$ and ΩN_c is interpreted as the number of sources per beam area. What is this value if q is taken to be 5?

Chapter 9
Interferometers and Aperture Synthesis

1. In one dimension, one can make a simple interferometer from a paraboloid by masking off all of the reflecting surface except for two regions of dimension a, which are separated by a length b, where $b >> a$. Assume that the power incident on these two regions is reflected without loss, then coherently received at the prime focus. A receiver there amplifies and square law detects these signals. This system is used to measure the response of an isolated source.

(a) In one dimension, one can make a simple interferometer from a paraboloid by masking off all of the reflecting surface except for two regions of dimension a, which are separated by a length b, where $b >> a$. Assume that the power incident on these two regions is reflected without loss, then coherently received at the prime focus. A receiver there amplifies and square law detects these signals. This system is used to measure the response of an isolated source. Write out a one-dimensional version of Eq. (6.60) from 'Tools', which is the far field pattern:

$$f(\mathbf{n}) = \frac{1}{2\pi} \int\!\!\!\int_{-\infty}^{\infty} g(\mathbf{x}')e^{-ik\,\mathbf{n}\cdot\mathbf{x}'}\, \frac{dx'}{\lambda}\frac{dy'}{\lambda}$$

Apply this equation and Eq. (6.61), which is the relation of the field pattern and the power pattern:

$$P_n = \frac{|f(\mathbf{n})|^2}{|f_{max}|^2}.$$

to determine the far-field pattern of this instrument.

© Springer International Publishing AG, part of Springer Nature 2018
T. L. Wilson, S. Hüttemeister, *Tools of Radio Astronomy – Problems and Solutions*, Astronomy and Astrophysics Library,
https://doi.org/10.1007/978-3-319-90820-5_9

(b) Use Eq. (9.6) from 'Tools' to analyze the response of such a one-dimensional 2 element interferometer consisting of 2 paraboloids of diameter a, separated by a distance b, measuring a star by a disk of size θ_s. Show how one can determine the angular size of the source from the response, R.

2. Show that the one-dimensional version of Eq. (9.6) from 'Tools', which is:

$$R(\mathbf{B}) = \iint_{\Omega} A(\mathbf{s}) I_\nu(\mathbf{s}) \exp\left[i\, 2\pi\nu\left(\frac{1}{c}\mathbf{B}\cdot\mathbf{s} - \tau_i\right)\right] d\Omega\, d\nu$$

is:

$$R(B) = \int A(\theta)\, I_\nu(\theta) \exp\left[i\, 2\pi\nu_0\left(\frac{B}{c}\right)\cdot\theta\right] d\theta$$

(this is Eq. (9.8) of 'Tools'; with $\theta/c = 1/\lambda$, it becomes Eq. (9.46) in 'Tools'). Interpret this relation in terms of the Fourier Transform (FT) pair with variables $u = B/\lambda$ and θ. In Fig. 9.4 of 'Tools', given here as figure 9.1, one dimensional distributions of $u = B/\lambda$ are shown. Show that the "Image-plane distributions" (on the right with variables x, $y=\theta_x$, θ_y) are related to the "one-dimensional" (u, v) plane distributions (on the left) using Eq. (9.46).

3. Use the result of problem 2 (which is Eq. (9.46) in 'Tools' and figure 9.2 here) to obtain the interferometer beam shapes (left side of the figure here, which is Fig. 9.18 in 'Tools') from the coverage of the (u, v) plane, shown on the right side of this figure.

4. This problem and the next two problems illustrate features of fig. 9.3 (shown above) and the use of Eq. (9.46), which is the result of problem 2.
The Sun is assumed to be a uniformly bright disk of diameter 30'. This source is measured using a multiplying interferometer at 10 GHz which consists of two identical 1 m diameter radio telescopes. Each of these dishes is uniformly illuminated. We assume that the instrumental phase is adjusted to zero, i.e. $\tau_i = 0$, the bandwidth of this system is small, and one measures the central fringe.
(a) What is the FWHP of each dish? Compare to the diameter of the Sun.
(b) Assume that the antenna efficiency and beam efficiency of each of the 1 m telescopes are 0.5 and 0.7, respectively. What is the antenna temperature of the Sun, as measured with each? What is the main beam brightness temperature measured with each telescope?
(c) Now the outputs are connected as a multiplying interferometer, with a separation on an east–west baseline of 100 m. Suppose the Sun is observed when directly overhead. What is the fringe spacing? Express the response in terms of brightness temperature measured with each dish individually.
(d) Now consider the more general case of a source which is not directly overhead. Determine the response as a function of B, the *baseline*.
(e) What is the response when the two antennas are brought together as close as possible, namely 2 m?

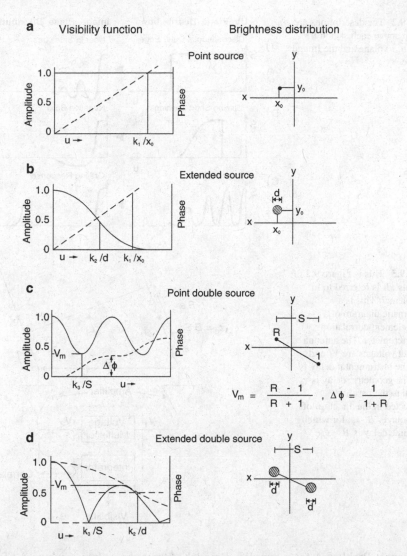

Fig. 9.1 This is Figure 9.4 in 'Tools', This shows the visibility function for various brightness distribution models. The solid lines are amplitudes, the dashed lines are phases. (**a**) A point source displaced from the phase center; a displacement of $x_0 = 1''$ shifts phase by one fringe for a $k_1 = 206,265$ wavelength baseline. (**b**) A Gaussian shaped extended source of FWHP $1''$ displaced from the origin; the amplitude reaches a value of 0.5 at $k_2 = 91,000$ wavelengths. (**c**) Two point sources with an intensity ratio R; the period of amplitude and phase depends on the separation. If the centroid of the double is the phase center, the sign of phase gives the direction of the more intense components, with positive to the east. (**d**) Two extended double sources; this has been obtained from the response to a pair of point sources by multiplying the visibility amplitude by the envelope shown in (**b**). The numerical values are $k_3 = 103,000$ if $s = 1''$ and $k_2 = 91,000$ if $d = 1''$ [after Fomalont and Wright (1974)]

Fig. 9.2 The descriptions are given above each sketch for the (u, v) plane and the Image Plane

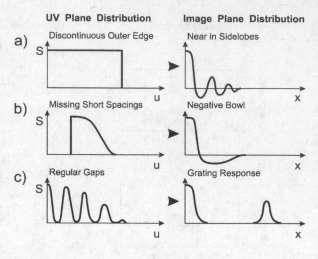

Fig. 9.3 This is Figure 9.2 of 'Tools'. It is referred to in problem 7. This is a schematic diagram of a two-element correlation interferometer. The antenna output voltages are V_1 and V_2; the instrumental delay is τ_i. The geometric delay is τ_g, equal to the baseline projected in the direction of the source, $B \cdot s$, for which the time delay is $B \cdot s/c$

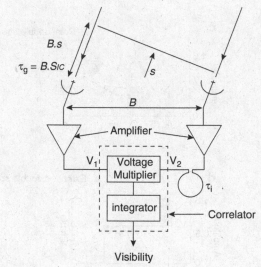

5. Repeat Problem 4 for a simplified model of the radio galaxy Cygnus A. Take this source to be a one-dimensional double with centers separated by $2\theta_1 = 1.5'$. Assume that each region have uniform intensity distributions, with FWHP sizes of $\theta_2 = 50''$. Each region has a total flux density of 50 Jy.

6. Repeat Problem 4 for the HII region Orion A, taking this as a one-dimensional Gaussian region with angular size FWHP 2.5'. Repeat for the supernova remnant Cassiopeia A, a ring-shaped source. In one dimension model this source is a region of outer diameter 5.5' with a ring thickness of 1'.

7. Suppose the receivers of an interferometer are double-sideband mixers. In each mixer, power arrives from the *upper sideband* and from the *lower sideband*. Use our figure 9.3 and Eq. (9.6) (see the beginning of problem 2) to show that the upper and lower sidebands can be separated since the geometric phase delays, τ_g for the upper and lower sideband frequencies $\phi_g = 2\pi \nu \tau_g$, will differ.

8. The interferometer described in Problem 4 is used to measure the positions of intense water masers at 22.235 GHz. The individual masers are very compact sources, unresolved even with interferometer antenna spacings of hundreds to thousands of kilometers. These masers normally appear as clusters of individual sources, but usually do *not* have identical, radial velocities.

(a) Discuss using a set of contiguous narrow frequency filters as a spectrometer. Should these filters be placed *before* or *after* multiplication? How wide a frequency band can be analyzed without diminishing the response of this system? What must the phase and frequency characteristics of these filters be?

(b)* An alternative to filters is a *cross-correlation* spectrometer. Discuss how this system differs from the filter system. Analyze the response of such a cross correlator system if the instrumental phase differences between antennas can be eliminated before the signals enter the cross correlator.

9. Suppose we use an interferometer for which one: (1) added the voltage outputs of the two antennas, and then square-law detected this voltage and (2) inserted a phase difference of 180° into one of the inputs, (3) repeated this process and (4) then subtracted these outputs to obtain the correlated voltages. Compare the noise arising from this process with that from a direct multiplication of the voltages. Show that direct multiplication is more sensitive.

10. Derive Eq. (9.12), which is:

$$R(B) = A\, I_0 \cdot \theta_0\, \exp\left(i\pi\, \frac{\theta_0}{\theta_b}\right) \left(\frac{\sin\left(\pi\theta_0/\theta_b\right)}{(\pi\theta_0/\theta_b)}\right)$$

using the one dimensional form of Eq. (9.46), from problem 2. Derive Eq. (9.13), which is:

$$\Delta\phi = 2\pi \left(\frac{\theta_{\text{offset}}}{\theta_b}\right)\left(\frac{\Delta\nu}{\nu}\right)$$

Show all steps in both derivations.

11. Use Eq. (9.27), which is: $\Delta T_B = \dfrac{2\, M\, \lambda^2\, T'_{\text{sys}}}{A_e \Omega_b \sqrt{2\, N t\, \Delta\nu}}$ setting $M = 1$, show that the following is an alternative form of this relation:

$$\Delta T_B \sim \frac{\lambda^{0.5}\, T_{\text{sys}}\, B^2_{\text{max}}}{n\, d^2 \sqrt{\tau \Delta V}},$$

where B_{max} is the maximum baseline of the interferometer system, d is the diameter of an individual antenna and ΔV is the velocity resolution. In addition, $N = n(n-1)/2 \approx n^2/2$ where n is the number of correlations.

12*. Suppose we have a filled aperture radio telescope with the same diameter and collecting area as an interferometer used to carry out a full synthesis.

(a) If the filled aperture diameter is $D(= B_{max}$ of problem 11) and the diameter of each individual interferometer antenna is d, how many elements are needed to make up the interferometer? (This is the number of dishes of diameter d which fit into the area of the filled aperture D.)

(b) Calculate the times needed to map a region of a given size with the filled aperture (equipped with a single receiver) and the interferometer array.

(c) The following is related to "mosaicing", which is needed for interferometer imaging of a very extended source of size Θ which is *very extended* compared to the beamsize of each individual interferometer antenna, θ. Calculate how many pointings are needed to provide a complete image of the extended source. The RMS noise for a map made with a single pointing is

$$\Delta T_B \sim \frac{1}{n\,d^2\sqrt{2\,N\,\tau\,\Delta\nu}}$$

(see previous problem). If the total time available for the measurement of a region is T, show that the number of pointings is proportional to T/d^2. Then show that the RMS noise in a mosaiced map is $\Delta T_B \sim 1/d$ instead of $\Delta T_B \sim 1/d^2$.

13. A source with a FWHP of $\sim 30''$ and maximum intensity of 2.3 K, T_{MB} is observed with ALMA interferometer. If a velocity resolution of $0.15\,\mathrm{km\,s^{-1}}$ is used to measure the $J = 1 - 0$ line of CO at 2.7 mm, with a $10''$ angular resolution, how long must one integrate to obtain a 5-to-1 peak signal-to-noise ratio? Use the result of problem 11.

14. The MERLIN interferometer system has a maximum baseline length of 227 km. At an observing frequency of 5 GHz, what is the angular resolution? Suppose that the RMS noise after a long integration is $50\,\mu\mathrm{Jy}$, that is, $5 \times 10^{-5}\,\mathrm{Jy}$. Use the Rayleigh–Jeans relation to obtain the RMS noise in terms of main beam brightness temperature. If a thermal source has at most a peak temperature of $5 \times 10^5\,\mathrm{K}$, can one detect thermal emission?

15. Compare the performance of the Jansky-VLA at 1.3 cm with the bilateral ALMA (i.e. fifty 12 m antennas) at 2.6 mm, for a velocity resolution of $0.1\,\mathrm{km\,s^{-1}}$. Set $M = 1$ for the Jansky-VLA. For the twenty-seven 25 m antennas of the JVLA 1.3 cm, each with $A_e = 240\,\mathrm{m^2}$, and a $3''$ synthesized beam. For the fifty 12 m diameter antennas of bilateral ALMA, each with $A_e = 80\,\mathrm{m^2}$ with a 140 K system noise temperature (including atmosphere) at 2.6 mm and a 38 kHz frequency resolution ($=0.1\,\mathrm{km\,s^{-1}}$). Use Eq. (9.29) from 'Tools':

$$\Delta T_B = 838.0 \frac{M\,\lambda^2\,T'_{sys}}{A_e\theta_B^2\sqrt{N\,t\,\Delta\nu}}$$

where T_B is in Kelvins, λ is in mm, θ_B in arc seconds, and $\Delta\nu$ in kHz.

Chapter 10
Emission Mechanisms of Continuous Radiation

1. Suppose an object of radius 100 m, with a uniform surface temperature of 100 K passes within 0.01 AU of the earth (an astronomical unit, AU, is 1.46×10^{13} cm).
(a) What is the flux density of this object at 1.3 mm?
(b) Suppose this object is observed with a 30 m telescope, at 1.3 mm, with a beamsize of 12″. Assume that the object has a Gaussian shape; calculate the peak brightness temperature by considering the dilution of the object in the telescope beam. Neglect the absorption by the earth's atmosphere.
(c) This telescope is equipped with a bolometer with NEP = 10^{-15} W Hz$^{-1/2}$ and bandwidth 20 GHz; how long must one integrate to detect this object with a 5 to 1 signal-to-noise ratio, if the beam efficiency is 0.5, and the earth's atmospheric optical depth can be neglected?

2. The Orion hot core is a molecular source with an average temperature of 160 K, angular size 10″, located 500 pc (= 1.5×10^{21} cm) from the Sun. The average local density of H_2 is 10^7 cm^{-3}.
(a) Calculate the line-of-sight depth of this region in pc, if this is taken to be the diameter.
(b) Calculate the column density, $N(H_2)$, which is the integral of density along the line of sight. Assume that the region is uniform.
(c) Obtain the flux density at 1.3 mm using $T_{dust} = 160$ K, the parameter $b = 3$, $\beta = 2$ and solar metallicity ($Z = Z_\odot$) in Eq. (10.6), which is:

$$N_H = 1.55 \times 10^{24} \frac{S_\nu}{\theta^2} \frac{\lambda^{2+\beta}}{Z/Z_\odot \, b \, T_d} \left(\frac{e^u - 1}{u} \right)$$

© Springer International Publishing AG, part of Springer Nature 2018
T. L. Wilson, S. Hüttemeister, *Tools of Radio Astronomy – Problems and Solutions*, Astronomy and Astrophysics Library,
https://doi.org/10.1007/978-3-319-90820-5_10

where S_ν is the flux density in mJy, the beam FWHP size, θ, is in arc seconds, the wavelength, λ in mm, T_d the dust temperature in Kelvins, $Z/Z_\odot$ is the fraction of the solar metallicity, (more generally, 'b' is an adjustable parameter with the value of 1.9 for moderate density and 3.4 for dense gas). [A skeptic might consider b and β to be fudge factors.] The value of 'u' is given by $\frac{14.4}{\lambda T} = \frac{0.048\,\nu}{T}$.

(d) Use the Rayleigh–Jeans relation (given in problem 10 of chapter 1) to obtain the actual dust continuum brightness temperature from this flux density, for a $10''$ source. Show that this is much smaller than T_{dust}.

(e) At long millimeter wavelengths, a number of observations have shown that the optical depth of such radiation is small. Then the observed temperature is $T = T_{\text{dust}} \times \tau_{\text{dust}}$, where the quantities on the right hand side of this equation are the dust temperature and dust optical depth. From this relation, determine τ_{dust}.

(f) At what wavelength is $\tau_{\text{dust}} = 1$ if $\tau_{\text{dust}} \sim \lambda^{-4}$?

3. (a) From figure 10.1 (also Fig. 10.1 in 'Tools'), determine the "turnover" frequency of the Orion A HII region, that is the frequency at which the flux density stops rising, and starts to decrease. This can be obtained by noting the frequency at which the linear extrapolation of the high and low frequency parts of the plot of flux density versus frequency meet. At this point, the optical depth, τ_{ff}, of free–free emission through the center of Orion A, is unity, that is $\tau_{\text{ff}} = 1$. Call this frequency ν_0.

(b) From Eq. (10.37), which is:

$$\tau_\nu = 8.235 \times 10^{-2} \left(\frac{T_e}{K}\right)^{-1.35} \left(\frac{\nu}{\text{GHz}}\right)^{-2.1} \left(\frac{\text{EM}}{\text{pc}\,\text{cm}^{-6}}\right) a(\nu, T)$$

where, usually, the parameter "a" is close to unity, gives the relation of turnover frequency, electron temperature, T_e, and emission measure. This relation applies to a uniform density, uniform temperature region; actual HII regions have gradients in both quantities, so this relation is at best only a first approximation. Determine EM for an electron temperature $T_e = 8300\,\text{K}$.

(c) The FWHP size of Orion A is $2.5'$, and Orion A is $500\,\text{pc}$ from the Sun. What is the linear diameter for the FWHP size? Combine the FWHP size and emission measure to obtain the RMS electron density.

4. A more accurate method to obtain the emission measure of the high electron density core of an HII region such as Orion A is to use $T_B = T_e \tau_{\text{ff}}$, where T_B is the brightness temperature of the source corrected for beam dilution.

(a) Use the T_e and source FWHP size values given in the last problem. For $\nu = 23\,\text{GHz}$, take the main beam brightness temperature, $T_{\text{MB}} = 24\,\text{K}$, and the FWHP beamsize as $43''$. Correct the main beam brightness temperature, T_{MB}, for source size to obtain T_B.

(b) Determine τ_{ff}.

(c) Use Eq. (10.37, as given in problem 3) with $a = 1$ to find ν_0 and EM; compare these results to those obtained in the last problem. Discuss the differences. Which

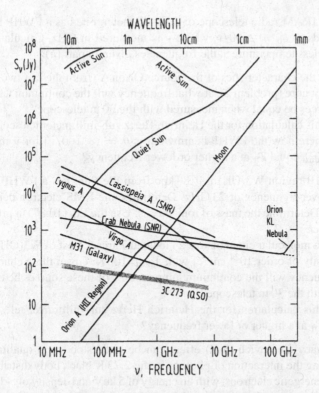

Fig. 10.1 This is Figure 1 in Chapter 10 of 'Tools'. It shows the flux densities of various radio sources. The Moon, the quiet Sun and (at lower frequencies) the H II region Orion A are examples of Black Bodies. At frequencies higher than 200 GHz there is additional emission from dust in the Orion KL molecular cloud. The active Sun, supernova remnants such as Cassiopeia A (3C461), the radio galaxies Cygnus A (3C405), Virgo A (Messier 87, 3C274) and the Quasi Stellar radio source (QSO) 3C273 are nonthermal emitters. The hatching around the spectrum of 3C273 is meant to indicate rapid time variability. (The 3C catalog is the third Cambridge catalog, a fundamental list of intense sources at 178 MHz (Bennett 1962 Mem RAS 68, 163 (the revised 3C catalog)); more recent continuum catalogs are Gregory et al. (1996 Ap. J. Suppl., 103, 427 the "GB6" survey), Condon et al. (1998 Astronom. J. 115, 1693, the "NVSS" survey) and Cohen et al. (2004, Ap. J. Suppl., 150, 417 the "VLSS" survey)

method is better for determining the EM value for the core of an HII region at high frequencies?

5. (a) For frequencies above 2 GHz, the optical depth of Orion A is small (i.e., the source is optically thin) and τ_{ff} varies as $v^{-2.1}$. Calculate τ_{ff} at 5, 10, 23, 90, 150 and 230 GHz.
(b) Next calculate the peak brightness temperature, at the same frequencies, for a telescope beam much smaller than the FWHP source size. Use the expression $T_B = T_e \tau_{ff}$.

(c) With the IRAM radio telescope of 30 m diameter, one has a FWHP beamwidth in arc seconds of $\theta_b = 2700/\nu$, where ν is measured in GHz. Calculate the main beam brightness temperature at the frequencies given in part **(a)**.

6. (a) Given the characteristics of the source Orion A (from the last two problems) and Orion hot core (problem 2), at what frequency will the continuum temperatures of these sources be equal when measured with the 30 m telescope?
(b) Repeat this calculation for the Heinrich Hertz sub-millimeter telescope, of 10 m diameter, where now the FWHP beamwidth is $\theta = (8, 100/\nu)$ for ν measured in GHz. Will T_{dust} equal T_{ff} at a higher or lower frequency?

7. (a) The HII region W3(OH) is 1.88 kpc from the Sun, has a FWHP size of $2''$ and a turn over frequency of 23 GHz. Determine the RMS electron density if the T_e=8500 K. Determine the mass of ionized gas. Use the Eq. (10.37) in problem 3(b) of this chapter.
(b) There is a molecular cloud of size $2''$ located 7 arcsec East of W3(OH). The dust column density of order 10^{24} cm^{-2}, with T_{dust}=100 K. Given these characteristics, at what frequency will the continuum temperatures of these sources be equal when measured with the 30 m telescope?
(c) Repeat this calculation for the Heinrich Hertz sub-millimeter telescope. Will T_{dust} equal T_{ff} at a higher or lower frequency?

8. The Sunyaev–Zeldovich (S-Z) effect can be understood in a qualitative sense by considering the interaction of photons in the 2.73 K black body distribution with much more energetic electrons, with an energy of 5 keV and density of $\sim 10^{-2}$ cm^{-3}.
(a) What is the energy of photons with a wavelength of 1.6 mm (the peak of the background distribution)? Compare to the energy of the electrons.
(b) Obtain the number of 2.73 K photons per cm^3 from Problem 3(d) of Chap. 11.
(c) Assume that the interaction of the 2.73 K black body photons with the electrons (assumed monoenergetic) in the cluster will lead to the equipartition of energy. Make a qualitative argument that this interaction leads to a net increase in the energy of the photons. Justify why there is a *decrease* in the temperature of the photon distribution for wavelengths longer than 1.6 mm and an *increase* shorter than this wavelength.

9. The source Cassiopeia A is associated with the remnant of a star which exploded about 330 years ago. Measurements of the radio emission give the relation of flux density to frequency, as shown in figure 10.1 in this chapter, and also Fig. 10.1 in 'Tools'. For the sake of simplicity, assume that the source has a constant temperature and density, in the shape of a ring, with thickness $1'$ and outer radius of angular size $5.5'$. What is the actual brightness temperature at 100 MHz, 1, 10, 100 GHz?

10. Obtain the integrated power and spectral index (power emitted per bandwidth proportional to ν^{-n}) for synchrotron radiation from an ensemble of electrons which have a distribution $N(E) = N_0$, that is a constant energy distribution from E_{min} to E_{max}.

Chapter 11
Some Examples of Thermal and Nonthermal Radio Sources

1. (a) This problem deals with the quiet Sun. The electron distribution is given by Eq. (11.2) in 'Tools', which is:

$$\frac{N_e}{cm^{-3}} = \left[1.55 \left(\frac{r}{r_0} \right)^{-6} + 2.99 \left(\frac{r}{r_0} \right)^{-16} \right] \times 10^8$$

with $r_0 = 7 \times 10^{10}$ cm to determine the emission measure, $\int N_e^2 dl$, of the quiescent solar atmosphere.

(b) Determine the optical depth of the quiescent solar atmosphere, looking at the center of the Sun, using Eq. (10.37) of 'Tools', which is stated in problem 3 (b) of chapter 10), with $T_e = 10^6$ K and for a frequency 100 MHz. What is the brightness temperature of the Sun?

2. If the nearest Jupiter is 10 pc from us, and are bursts which are identical with the Jupiter bursts, i.e. with an intrinsic peak intensity of 10^5 Jy at the distance of Jupiter (mean= 778 million km), what is the flux density from such a burst at 10 pc?

3. (a) At what frequency does the intensity of a 2.73 K black body reach a maximum? At what wavelength? Use the $T - \lambda$ relation in Eq. (1.26), which is:

$$\left(\frac{\lambda_{max}}{cm} \right) \left(\frac{T}{K} \right) = 0.28978 .$$

and the $T - \nu$ relation in Eq. (1.25), which is:

$$\left(\frac{\nu_{max}}{GHz} \right) = 58.789 \left(\frac{T}{K} \right)$$

© Springer International Publishing AG, part of Springer Nature 2018
T. L. Wilson, S. Hüttemeister, *Tools of Radio Astronomy – Problems and Solutions*, Astronomy and Astrophysics Library,
https://doi.org/10.1007/978-3-319-90820-5_11

(b) Could the difference between the maximum wavelength and frequency be caused by the different weightings of the Planck relation? Determine the intensity, I_ν, which in this case is the Planck function, B_ν, at the maximum frequency. B_ν is defined in problem 14 of chapter 1.

(c) What is the (integrated) energy density, given by: $u = (1/c) \int I d\Omega = (4\pi/c) I$?

(d*) Reformulate the derivation of the Stefan–Boltzmann relation in 'Tools' Eq. (1.24), which is:

$$B(T) = \sigma T^4, \qquad \sigma = \frac{2\pi^4 k^4}{15 c^2 h^3} = 1.8047 \times 10^{-5} \, \text{erg cm}^{-2} \, \text{s}^{-1} \, \text{K}^{-4}$$

to obtain the number density of photons. To do this, one must integrate the Planck function (Eq. (1.13)):

$$B_\nu(T) = \frac{2 h \nu^3}{c^2} \frac{1}{e^{h\nu/kT} - 1}$$

this function is divided by $h\nu$ and then over all frequencies. Make use of the relation with $x = h\nu/kT$:

$$\int_0^\infty \frac{x^2}{e^x - 1} dx = 2.404$$

to determine how many photons are present in a volume of $1 \, \text{cm}^{-3}$.

(e) What is the error in applying the Rayleigh–Jeans approximation, instead of the Planck relation to calculate the intensity of the 2.73 K black body radiation at 4.8, 115 and 180 GHz? Compare these values to Fig. 1.7 of 'Tools', given here as figure 11.1.

4. The 2.73 K microwave background (a black body) is one of the most important pieces of evidence in support of the big bang theory. The expansion of the universe is characterized by the red shift z. The ratio of the observed wavelength, λ_o, to the (laboratory) rest wavelength, λ_r, is related to z by $z = (\lambda_o/\lambda_r) - 1$. The dependence of the temperature of the 2.73 K microwave background on z is $T = 2.73\,(1 + z)$. What is the value of T at $z = 2.28$? What is the value at $z = 5$ and $z = 1000$?

5. Use Eq. (1.25), which is:

$$\left(\frac{\nu_{max}}{\text{GHz}} \right) = 58.789 \left(\frac{T}{\text{K}} \right)$$

together with Fig. 1.7 of 'Tools' (figure 11.1 here) to repeat the analysis of problem 3. How do the results compare?

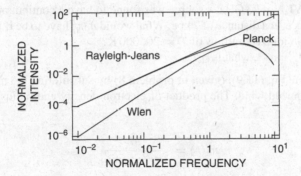

Fig. 11.1 This is Figure 1.7 from 'Tools', used in problem 3(e). This is the normalized Planck curve and the Rayleigh-Jeans and Wien approximation. The 2.73 K background follows a Planck relation, with a peak at 160 GHz. This is at the normalized frequency of 3. Thus for frequencies below 50 GHz, the Rayleigh-Jeans approximation is adequate for most purposes. For frequencies larger than 160 GHz, the Wien curve provides a reasonable approximation

6. (a) At 5 GHz, the brightness temperature in the outer parts of Orion A is $\sim$0.5 K. Use the assumption of an optically thin, smooth Bremsstrahlung emission from a region with $T_e = 6500$ K which fills the telescope beam completely to calculate the brightness temperature of these regions at 23 GHz. Use $T_B = T_e \times \tau$ with τ from problem 3 of chapter 10. At 23 GHz the map of Orion A has an RMS noise of 0.1 K. Would this emission from the outer parts of Orion A be detected in the 23 GHz map? **(b)** At what frequency would the outer regions of Orion A have an optical depth of unity?

7. Suppose a solar type star is to be detected at the 1 μJy level at $\lambda = 3$ mm. Given that $T_e = 5700$ K and $r_0 = 7 \times 10^{10}$ cm, what is the maximum distance that such a star can be detected? use the relation in Eq. (11.1), which is:

$$\left(\frac{S}{\mu Jy}\right) = 7 \left(\frac{R_*}{R_\odot}\right) \times \left(\frac{1}{D(pc)}\right) \times \left(\frac{\nu(GHz)}{345\,GHz}\right)$$

8. (a) Calculate the radio continuum flux density at $\nu = 10$ GHz for a B3 supergiant $(T = 1.6 \times 10^4$ K, $r_0 = 3.6 \times 10^{12}$ cm). Use an electron and ion density of 10^{10} cm^{-3} and Eq. (11.9) of 'Tools', which is:

$$S_\nu = 8.2 \left(\frac{n_0\, r_0^2}{10^{36}}\right)^{4/3} \left(\frac{\nu}{GHz}\right)^{0.6} \left(\frac{T_e}{10^4\,K}\right)^{0.1} \left(\frac{d}{kpc}\right)^{-2}$$

with $r = r_0$ for such a star which is 3 kpc distant.
(b) Is this source detectable with the 100 m telescope if the receiver noise is 50 K, if 1 Jy corresponds to 1.3 K, T_A, and the receiver bandwidth is 500 MHz? Do not consider confusion effects.

(c) With the JVLA at 23 GHz, a source was found to have a continuum flux density of 27 mJy. This is at a distance of 7 kpc. What would $n_0 r_0^2$ have to be if this emission be caused by an ionized outflow of $T = 20,000$ K?

(d) If $n_0 = 10^{10}$ cm^{-3}, what is r_0?

9*. Reformulate Eq. (11.9), (given in problem 8) by substituting the mass loss rate for a steady ionized wind. The product of electron density and radius squared can be related to

$$n_e(r) = \frac{\dot{M}}{4\pi r^2 v_w \mu m_H} \ .$$

Substitute this relation into the equation in Problem 4, where $\dot{M}$ is the mass loss rate in $10^{-5}\ \dot{M}$ (per year) and v_w is the wind velocity, in units of 1000 km s^{-1}. μ is the average mass as a multiple of the mass of the hydrogen atom, m_H. Use this result to find the mass loss rate for the source analyzed in Problem 8.

(a) if $v_w = 100$ km s^{-1}.

10. The parameters of a B0 Zero Age Main Sequence (ZAMS) star are $T = 3.1 \times 10^4$ K, luminosity $L = 2.5 \times 10^4\ L_\odot$ and radius $r = 3.8 \times 10^{11}$ cm. Suppose this object has a mass loss rate of $10^{-6} M_\odot$ per year and is 7 kpc distant. What is the flux density for a frequency of 10 GHz? Is this source detectable with the 100 m telescope? With the JVLA?

11. From the flux density at 100 MHz in Fig. 10.1 of 'Tools' (figure 10.1, chapter 10), calculate the peak brightness temperature of the Crab nebula, if the FWHP angular size of this source is $5'$, and the source shape is taken to be Gaussian. Repeat this calculation for a frequency of 10 GHz, using the same angular size. If the maximum brightness temperature for Bremsstrahlung emission from a pure hydrogen HII region is considered to be 20,000 K, is the emission from the Crab nebula thermal or non-thermal?

12. If Cassiopeia A has an angular diameter of $5.5'$, determine the present-day linear size of Cassiopeia A if this source is 3 kpc from the Sun. If the explosion occurred in 1667, and if the expansion velocity has been constant, what is v_{exp}? The JVLA can measure positions of "point" (i.e. unresolved in angle) features in Cas A, and if these features do not change shape with time, but merely move with v_{exp}, over what time scale would you have to carry out JVLA measurements to observe expansion?

13. (a) Use Eq. (11.45) of 'Tools', which is:

$$\frac{\dot{S}_\nu}{S_\nu} = -\frac{4}{5}\frac{\delta}{t}$$

with $\delta = 2.54$, to extrapolate the radio flux density of Cassiopeia A to a time when this source was 100 years old; that is, what was S_ν at 100 MHz in 1777? See Fig. 10.1 of 'Tools' (also figure 10.1 in chapter 10) for the flux today. What would be the angular size if the expansion is linear with time?

(b) Calculate the peak brightness temperature in 1777 assuming that this source is a Gaussian, using Eq. (8.19) of 'Tools', which is:

$$S = 2.65 \frac{T_{MB} \, \theta_o^2}{\lambda^2}$$

14. There is a sharp decrease in the flux density of Cassiopeia A at a frequency of about 10 MHz. If this source is 3 kpc from the Sun, and the average electron density is $0.03 \, \text{cm}^{-3}$, calculate whether the cause of the fall off is free–free absorption by electrons along the line of sight. These will have an effect only if $\tau = 1$. Use Eq. (10.35 of 'Tools', given in problem 3 (b) of chapter 10 in this volume) with $T_e = 6000 \, \text{K}$.

15. (a) Make use of the *minimum energy theorem* to estimate the magnetic fields and relativistic particle energies on the basis of synchrotron emission, using Eq. (10.121), which is:

$$W_{tot} = \frac{7}{4}(6\pi)^{-3/7} \left(\frac{G\nu^n}{H} S_\nu \right)^{4/7} R^{8/7} \, V^{3/7}$$

to obtain a numerical result if the spectral index, n, is 0.75, and $b(n) = 0.086$. For the maximum frequency, take ν_{max} equal to 50 GHz and for the minimum frequency, ν_{min}, equal to 0.1 GHz. Finally, take η (the ratio of other relativistic particles to that of electrons) to be 10. With these parameters, show that the expression for the B field is

$$B_{eq} = 1.2 \times 10^{-5} \left(\frac{S_\nu[\text{Jy}] \, R^2[\text{Mpc}] \nu^{0.75}[\text{GHz}]}{V[\text{kpc}]} \right)^{2/7} .$$

16. Assume that the galaxy NGC 253 is similar to our Milky Way. The radius of the synchrotron-emitting halo is 10 kpc at a distance of 3.4 Mpc. At $\nu = 8.7$ GHz, the integrated flux density is 2.1 Jy and the spectral index is $n = 0.75$ ($S_\nu = S_0(\nu/\nu_0)^{-0.75}$). Take $\nu_{max} = 50$ GHz and $\nu_{min} = 10$ MHz to calculate the B field and estimate the relativistic particle energy assuming that the minimum energy condition holds, i.e. using Eq. (10.121 of 'Tools' as given in problem 15 of this chapter).

17. Assume that the distance to Cygnus A is 170 Mpc. This source has a flux density of 10^4 Jy at 100 MHz. Assume that the electrons radiate over a frequency range from 10 MHz to 50 GHz with a spectral index $n = 0.75$. Find the power, P, radiated by the electrons via the synchrotron process, using

$$P = 4\pi R^2 \int_{\nu_{min}}^{\nu_{max}} S_\nu d\nu .$$

Compare to the total energy of the radio lobes, 2×10^{57} erg, calculated under the assumption of equipartition. What is the lifetime of these relativistic electrons if synchrotron emission is the only loss mechanism? Compare this to the expected lifetime of the source if the lobes are 7×10^4 pc apart and are thought to be moving with a speed $<0.2c$. What do you conclude about the need to replenish the energy of the electrons?

18. (a) The quasi-stellar radio source 3C273 has a red shift of 0.16. Take the Hubble constant, H_0, to be $70 \, \mathrm{km \, s^{-1}}$ per Mpc. Find this distance. The flux density varies on a time scale of months. Use a simple relation of $R = c t$ to determine the source size, without taking any relativistic effects into account. What is the angular size? Next, using this angular size, convert the flux density at 20 GHz, which is $\sim$20 Jy, into a source brightness temperature. What is the result? Does this exceed the maximum temperature of 10^{12} K, the limit predicted by the inverse Compton effect? This is an indication that relativistic beaming effects are important.
(b) Make use of Eq. (11.61) of 'Tools', which is:

$$ v_{\mathrm{app}} = \frac{r}{t_2 - t_1} = \frac{V}{1 - \dfrac{V}{c} \cos \theta} $$

for transverse velocities this is given by Eq. (11.62) of 'Tools', which is:

$$ (v_{\mathrm{app}})_{\mathrm{tranverse}} = \frac{V \sin \theta}{1 - \dfrac{V}{c} \cos \theta} $$

What is the angle at which the apparent transverse velocity is a maximum? What is the apparent velocity at this angle? If the apparent expansion velocity is $7c$, what is the beaming angle? There is no counter jet. Explain why not, taking "Doppler boosting" into account, using Eq. (11.68) of 'Tools', which is:

$$ S(v) = \frac{L_0(v(1+z))}{4\pi R^2 (1+z)^3} $$

Chapter 12
Spectral Line Fundamentals

1. Use Eq. (12.4) of 'Tools' which is

$$\frac{n_2}{n_1} = \frac{g_2}{g_1} \exp(-\frac{h\nu}{kT})$$

to estimate T for a two-state system with equal statistical weight factors and level populations $n_1 = 1.01 n_2$ (where the upper state is n_2). Repeat for $n_1 = 1.1 n_2$. For non-Local Thermodynamic Equilibrium (LTE) conditions, T is referred to as T_{ex}.

2. We now investigate the variation of T_{ex} with the collision rate, C_{21}, and the spontaneous decay rate, A_{21}, for a two-level system., for $T_K \gg T_0 = h\nu/k$ is given by Eq. (12.41), which is:

$$T_{ex} = T_K \left(\frac{T_0 C_{21} + T_b A_{21}}{T_0 C_{21} + T_K A_{21}} \right)$$

where ν_{ul} is in GHz and μ_{ul} in 10^{-18} ESU units, or Debyes. The relation for collision rate gives the dependence on the kinetic temperature, T_K, the temperature of the radiation field, T_b, and the ratio of collision rates to A coefficients. In addition, we suppose that the collision rate between levels 2 and 1, C_{21}, is given by $n \langle \sigma v \rangle$, where the value of $\langle \sigma v \rangle$ is $\sim 10^{-10}$. The value of T_0 is $h\nu/k$ where h is Planck's constant, k is Boltzmann's constant and ν is the line frequency. When $n \langle \sigma v \rangle = A_{21}$ for the transition involved, this is referred to as the "critical density", n^*. For the 21 cm line, $A_{21} = 2.85 \times 10^{-15} \, \text{s}^{-1}$. Find n^* for this transition. For neutral hydrogen, in most cases, only two levels are involved in the formation and excitation of the 21 cm

© Springer International Publishing AG, part of Springer Nature 2018
T. L. Wilson, S. Hüttemeister, *Tools of Radio Astronomy – Problems and Solutions*, Astronomy and Astrophysics Library,
https://doi.org/10.1007/978-3-319-90820-5_12

line since the $N = 2$ level is 9 eV higher. Less secure is any result for multi-level systems. However, to obtain an order of magnitude estimate, repeat this calculation for the $J = 1 - 0$ transition of the molecule HCO^+, modelling the molecule as a two-level system in which the Einstein A coefficient is $A_{21} = 3 \times 10^{-5}\,s^{-1}$. What is the value of n^*? Compare this to the value for the 21 cm line. For HCO^+, take $T_K = 100\,K$; find the value of the local density for which $T_{ex} = 3.5\,K$. $T_b = 2.7\,K$. For the same density, calculate n^* for the $J = 1 - 0$ transition of the carbon monoxide molecule, CO, modelling this as a two-level system with $A_{21} = 7.4 \times 10^{-8}\,s^{-1}$.

3*. Line shapes can be obtained using a semi-classical model of the atom. Use the model of a classical oscillator, but now with a loss term proportional to velocity: $\ddot{x} = -\omega_0^2 x - \gamma \dot{x}$.

(a) Solve for x under the assumption that $\gamma \ll \omega_0$, using $x = x_0 e^{\alpha t}$.

(b) Determine the electric field caused by the motion of the oscillating charge.

(c) Determine the line shape using the Fourier transform (F.T.) of the electric field.

(d) Obtain the line intensity as the absolute value of the square of the F.T. of the electric field. This is the *Lorentzian* line shape.

(e) Determine the line shape if the thermal motion of the atoms, described by $f(v) = (m/2kT)^{3/2} \exp(-mv^2/2kT)$, is combined with the relation for the Doppler shift, $\Delta v/c = \Delta \nu/\nu_0$.

(f) Assume that the areas of these two line profiles are equal, and plot the line shapes. Discuss the difference in intensities of the line wings.

(g) Compare values of γ and ω_0 for the Lyman α line, given that the line frequency, ν, is $3.29 \times 10^{15}\,s^{-1}$ and the A coefficient is $5.4 \times 10^9\,s^{-1}$. Take γ as A, the Einstein coefficient for spontaneous decay. Repeat this for the 1.420 GHz line of hydrogen, emitted by hydrogen atoms in regions of density $1\,cm^{-3}$, $10^5\,cm^{-3}$ and $10^{19}\,cm^{-3}$ if $\gamma = 2.87 \times 10^{15}\,s^{-1}$.

4*. The energy of the ground state of the hydrogen atom can be obtained using the following analysis, which is closer to the spirit of quantum mechanics than the usual semi-classical orbit analysis. Assume that the nucleus has a very large mass, and charge e. The electron has a mass m and change $-e$. The electron moves with a momentum p at a distance x from the nucleus.

(a) Write down the energy equation for this situation.

(b) Use the relation obtained in problem 4 in chapter 2, namely $\Delta x \Delta k = 1$. Use the de Broglie relation $k = p/\hbar$ in the energy equation. Differentiate the energy equation, and set the result to zero to obtain the minimum value of x. What is this value? Compare to the lowest Bohr orbit. Calculate the energy. The value x is the lowest orbit of the electron. The radius increases with n^2, where n is the principal quantum number. Calculate the energy of the lowest two orbits. Now take the difference and set the energy difference equal to $h\nu$. What is the value of ν? Compare this to the frequency of the Lyman α line.

5. Evaluate the constants in Eq. (12.24) of 'Tools', which is:

$$A_{mn} = \frac{64\pi^4}{3\,h\,c^3}\, \nu_{mn}^3\, |\mu_{mn}^*|^2$$

to show that

$$A_{ul} = 1.165 \times 10^{-11} \times \nu_{ul}^3 |\mu_{ul}|^2 \qquad (12.1)$$

where ν_{ul} is in GHz, and μ_{ul} is in 10^{-18} esu units.

Chapter 13
Line Radiation of Neutral Hydrogen

1. The ratio of the populations in the upper, N_u and lower, N_l levels of the ground state of HI is given by the Boltzmann relation, where the statistical weights in these levels are 3 and 1, respectively: $N_u/N_l = 3\exp(-0.0682/T_s)$. Assume that the spin temperature, T_s, equals the kinetic temperature, T_K. Calculate the population ratio for a temperature of 100 K. Repeat for a temperature of 3 K (the lowest temperature possible under local thermodynamic equilibrium), for 10^4 K (the warm interstellar medium) and 10^6 K. Compare the differences in populations.

2. In this problem, we determine the value of the FWHP linewidth in terms of T_K, the kinetic temperature, and the mass of the emitter, m. We assume that the thermal motion of atoms in three dimensions is described by the Boltzmann relation for velocities v between $\pm\infty$: $f(v) = (m/2\pi k T_K)^{3/2}\exp\left(-\frac{mV^2}{2kT_K}\right)$.

(a) Show that the requirement $\int f(v)\mathrm{d}V = 1$ is fulfilled.

(b) Use the distribution from part **(a)**, with the definition

$$V_{RMS} = \sqrt{\int V^2 f(v)\mathrm{d}V}.$$

The above relation between the line-of-sight FWHP $\Delta V_{1/2}$ (which is measured) and the three-dimensional V_{rms} is $\Delta V_{1/2} = \sqrt{8\ln2/3}\, V_{RMS}$. Use this to relate the measured linewidth to the emitter mass and kinetic temperature for hydrogen. Show that $1\,\mathrm{km\,s^{-1}}$ is equivalent to motion in a gas of $T_K = 21.2\,\mathrm{K}$ for atomic hydrogen.

(c) Show that the general result is $T_K = 21.2\,(m/m_H) \times (\Delta V_{1/2})^2$, where m_H is the mass of a hydrogen atom.

(d) Compare this value with the speed of sound in an isothermal gas, $c_0 = \sqrt{P/\rho}$, where P is pressure in $\mathrm{dyne\,cm^{-2}}$ and ρ is density in $\mathrm{g\,cm^{-3}}$. Check that this is dimensionally correct, then evaluate c_0 in terms of kinetic temperature and density (in $\mathrm{cm^{-3}}$) for a perfect gas consisting of hydrogen atoms.

© Springer International Publishing AG, part of Springer Nature 2018
T. L. Wilson, S. Hüttemeister, *Tools of Radio Astronomy – Problems
and Solutions*, Astronomy and Astrophysics Library,
https://doi.org/10.1007/978-3-319-90820-5_13

Fig. 13.1 This is Fig. 13.9 of
'Tools'. It shows the
excitation cross section (in
arbitrary units) for a two level
system in which the energy
levels are separated by E_0

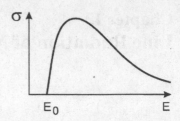

Table 13.1 Parameters of some selected atomic lines

Element and ionization state	Transition	ν/GHz	A_{ij}/s^{-1}	Critical density n^*	Notes
DI	$^2S_{1/2},\ F = 3/2 - 1/2$	0.327	4.65×10^{-17}	~ 1	a,b
HI	$^2S_{1/2},\ F = 1 - 0$	1.420	2.87×10^{-15}	~ 1	a,b
$^3\text{He}^+$	$^2S_{1/2},\ F = 0 - 1$	8.665	1.95×10^{-12}	~ 10	a

[a] Ions or electrons as collision partners
[b] H_2 as a collision partner

3*. In fig. 13.1 given here, we show the (idealized) cross section for a neutral–neutral collision to excite the population of a two-level system (levels are separated by an energy E_0). On the basis of this description, explain the behavior of the cross section with particle energy.

4*. For the electronic ground state ($L = 0$) of HI, DI and $^3\text{He}^+$, the energy of interaction of the electron and nuclear magnetic moments is given by

$$W = \frac{4}{3}\mu_\text{e} \times \mu_\text{n}\left(\frac{4Z^3}{I n^3}\right)$$

where Z is the nuclear charge, I is the angular momentum quantum number of the nucleus, and n is the principal quantum number. For HI, the energy levels are designated by the quantum numbers $F_\text{u} = 1$, $F_\text{l} = 0$ and $I = 1/2$. For DI, $F_\text{u} = 3/2$, $F_\text{l} = 1/2$ and $I = 1$. In both cases, $J = 1/2$. The magnetic moment of the HI nucleus is 2.79 nuclear magnetons. For the DI nucleus this is 0.857 nuclear magnetons. Use the frequency in Table 13.1 (we include the relevant portion of this Table) to scale the DI frequency from the HI frequency.

5. Repeat the previous problem for the hyperfine interaction for $^3\text{He}^+$. For this ion, $Z = 2$, and the upper and lower energy levels have quantum numbers $F_\text{u} = 0$ and $F_\text{l} = 1$. The magnetic moment of the $^3\text{He}^+$ nucleus is -2.1274 nuclear magnetons.

6. An estimate of $|\mu_\text{ul}|$ for hyperfine transitions is: $|\mu_{ul}^2| = \beta^2\mu_\text{e}^2$ where β^2 is 4/3 for a spin -1 system (D nucleus), and 1 for a spin $-1/2$ system (H nucleus). For HI, $A_{ul} = 2.869 \times 10^{-15}\,\text{s}^{-1}$. Given this result, the relations above, and the result of problem 5, chapter 12, use the line rest frequencies of HI, DI and ^3He, 1420.406 MHz, 327.384 MHz and 8665.65 MHz, respectively, and scaling arguments to obtain the A coefficients for DI and $^{3}He^{+}$. Compare the results with the compilation in Table 13.1.

7*. This problem outlines an estimate of the amount of telescope integration time needed to detect the 92 cm DI line toward an intense background continuum source. When the 100 m telescope, FWHP beam size 9' at 21 cm, and beam efficiency, $\eta_B = 0.75$ is used to measure absorption of the 21 cm line of HI toward the supernova remnant Cassiopeia A, one finds an apparent optical depth, $\tau_{app} = -\ln(1 - (T_{line})/(T_{cont})$, of 2.5. The total continuum flux density of Cas A at 21 cm is $S = 3000$ Jy; this varies with wavelength as $S \sim \lambda^{0.7}$. Take the FWHP size of Cas A as 5.5' and assume that the source and beam are Gaussian shaped. We want to search for deuterium using the hyperfine transition.

(a) Use the source and telescope parameters to estimate the FWHP beamwidth of the 100 m telescope, and the peak continuum antenna temperature at the DI wavelength, 92 cm.

(b) If the HI and DI lines arise from the same region, the linewidths in km s^{-1} are equal. The linewidth in frequency units, Δv_l, will follow the Doppler relation $(\Delta V/c = \Delta v_l/v_l)$. The DI profile is assumed to have a FWHP of 2 km s^{-1}; estimate the FWHP of the DI line profile in kHz.

(c) Use Eq. (12.17), which is:

$$\kappa_v = \frac{c^2}{8\pi} \frac{1}{v_0^2} \frac{g_2}{g_1} N_1 A_{21} \left[1 - \exp\left(-\frac{hv_0}{kT}\right) \right] \varphi(v)$$

and set $\varphi(v)$, the linewidth distribution, equal to dV obtain an expression for the *total* column density of DI. The relation of populations in the upper and lower energy levels is described in problem 1 of chapter 12. The general relation between excitation temperature, line optical depth and column density, $N = n \times L$, of *any* two-level system, is:

$$N_1 = 93.5 \frac{g_1 v^3}{g_u A_{ul}} \frac{1}{\left[1 - \exp(-4.80 \times 10^{-2} v/T_{ex})\right]} \int \tau dV , \qquad (13.1)$$

where T_{ex} is the excitation temperature, defined in problem 1 of chapter 12. N_1 is the column density in the lower level (in cm^{-2}, τ is the optical depth of the spectral line, dV is the linewidth, and v is the frequency of the line in GHz.

(d) Derive this relation.

(e) For DI, as for HI, $T_{ex} = T_s = T_{kin}$. Assume that $hv \ll kT_{ex}$ to simplify this relation. For the 92 cm line, $A_{ul} = 4.63 \times 10^{-17}$ s^{-1}, $g_u = 4$ and $g_l = 2$, what is the relation for DI?

Assume that the spin temperatures, T_s, of DI and HI are equal that $\tau(HI) = 2.5$ and that the D/H ratio is 1.5×10^{-5}. What is the antenna temperature of the DI line at 92 cm if $T_{line} = T_{cont}\tau$? From the DI line antenna temperature and the system noise (= receiver noise of 100 K plus the source noise), determine the integration time needed to detect a DI line, for a spectral resolution which is 1/2 of the FWHP linewidth. Compare to the results in Heiles et al. 1993 ApJ Suppl **89**, 271.

[Later searches and detections by Chengalur et al. 1997 A & A, 1997 **318**, L35 and Rogers et al. 2007 AJ **133**, 1625) were carried out toward positions in the outer galaxy. These outer galaxy searches are more favorable since the system noise is lower and the abundance of deuterium should be larger, due to the effect of stellar destruction of deuterium (see Wilson & Rood 1994 Ann. Rev. A & A 32, 191). The Chengalur et al. result was at the 3σ level; the Rogers et al. result gave a 9σ detection].

8. **(a)** Suppose a uniform, extended HI cloud has a physical temperature of $T_K = 2.73$ K. If the only background source is the 2.73 K microwave background, would you expect to observe the HI line in emission or absorption or no line radiation at all?

(b) Repeat if there is a background source with main beam brightness temperature, $T_{MB} = 3$ K, that is, $T_{MB} > T_K$. What would be the temperature of the absorption, ΔT_L, in K if $\tau = 1$?

   (c) Repeat for $T_K = 3.5$ K.

9. In this and the next two problems, we investigate the details of geometry (Fig. 13.10 in 'Tools' and figure 13.2 here). Assume that all regions are Gaussian shaped. There is a continuum source behind the cloud of neutral gas containing HI. The HI in this cloud has an excitation temperature T_{cl} and an angular size θ_{cl}. The continuum source has an *actual* brightness temperature T_0 and angular size θ_0. Assume that θ_a, the beam size of the antenna, is much larger than all other sizes. The cloud covers a fraction f of the background source, that is $\theta_{cl} = f\theta_0$. Specify the conditions under which there will be line absorption against the continuum source. Obtain the expression for the main beam brightness temperature of the line, $\Delta T_L = T_L - T_0$. If $\tau \ll 1$, show that $|\Delta T_L/T_0| = f\tau$.

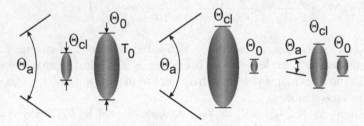

Fig. 13.2 This is Figure 10 in Chapter 13 of 'Tools'. This is a set of three sketches for problems 9, 10 and 11. These deal with the geometry of a neutral gas cloud in front of a continuum source where the relative sizes of the antenna beam, θ_a, the cloud, θ_{cl} and the continuum source θ_0, differ

10. Repeat the last problem for the situation in which $\theta_{cl} \gg \theta_0$, but *both* are much smaller than the antenna beam, θ_a. Obtain the expression for the main beam brightness temperature. Under what conditions does one find line absorption?

11. Repeat for the case in which the antenna beam is much smaller than either θ_{cl} or θ_0. Under what conditions does one find absorption?

Chapter 14
Recombination Lines

1*. A spherically symmetric, uniform HII region is ionized by an O7 star (mass about 50 $M_\odot$), with an excitation parameter, which is defined as $U = N_e^{2/3} L(\text{pc})$ has $U = 68 \, \text{pc} \, \text{cm}^{2/3}$. Background information needed: The excitation parameter U is used in the following problem. U is the radius of the HII region (in pc), multiplied by the 2/3 power of the electron density.
(a) Interpret the meaning and limitations of the excitation parameter. (See Table 14.1 in 'Tools'; the most relevant data for the spectral types of the exciting stars are given in the problems in this chapter)
(b) If $N_e = 10^4 \, \text{cm}^{-3}$ what is the radius of this region?
(c) Calculate the *Emission Measure*, EM $= N_e^2 L$, where L is the diameter of the HII region.
(d) If this region consists of pure hydrogen, determine the mass.

2*. If the ionization is caused by a cluster of B0 stars (each with mass 18 $M_\odot$), each with $U = 24 \, \text{pc} \, \text{cm}^{2/3}$, how many of these stars are needed to provide the same excitation as with one O7 star?

3*(a). Compare the mass of the HII region in Problem 1 to that of the exciting stars needed to ionize the regions in Problems 1 and 2.
(b) Suppose that the HII region in Problem 1 has an electron density, N_e, of $3 \times 10^4 \, \text{cm}^{-3}$, but the same Emission Measure, EM $= N_e^2 L$, where L is the diameter of the HII region. Determine the mass of ionized gas in this case.
(c) Now repeat this calculation for the same excitation parameter, but with $N_e = 3 \times 10^3 \, \text{cm}^{-3}$.

4. In the core of the HII region Orion A, the diameter is 0.54 pc, the emission measure, $N_e^2 L = 4 \times 10^6 \, \text{cm}^{-6} \, \text{pc}$, and the electron density N_e is $10^4 \, \text{cm}^{-3}$ (from optical data). Combine N_e with the emission measure to obtain the line-of-sight depth. Compare this result with the RMS electron density obtained by assuming a spherical region with a line-of-sight depth equal to the diameter. The "clumping

© Springer International Publishing AG, part of Springer Nature 2018
T. L. Wilson, S. Hüttemeister, *Tools of Radio Astronomy – Problems and Solutions*, Astronomy and Astrophysics Library,
https://doi.org/10.1007/978-3-319-90820-5_14

Fig. 14.1 This is Fig. 1 in Chapter 14 of 'Tools'. This shows the photoionization cross sections for H^0, He^0 and He^+ (Osterbrock & Ferland 2006 "Astrophysics of Gaseous Nebulae and Active Galactic Nuclei 2^{ed}" University Science Books, Herndon VA)

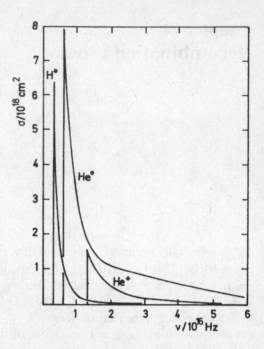

factor" is defined as the ratio of the actual to the RMS electron densities. What is this factor?

5. The assumption in problem 2 is that all of the exciting stars are of the same spectral type (and mass). This is not found to be the case. Rather the distribution of stellar masses follows some distribution. One is the *Salpeter* distribution, $N(M) = N_0 M^{-1.35}$. Integrate over mass $\int M N(M) dM$ to obtain the total mass of stars between the limits M_{lower} and M_{upper}. Take M_{lower} as 0.08 $M_\odot$, and M_{upper} as 50 $M_\odot$. Is there more mass in stars of type B0 and larger or in stars with masses below class B0?

6*. (a) In Fig. 14.1 in 'Tools' and here (fig. 14.1), is a sketch of the photoionization cross sections for hydrogen and two ionization states of helium. Explain why there is a sharp decrease in the absorption cross section for frequencies lower than ν_0. Calculate the photon energy corresponding to ν_0.

(b) At frequencies higher than ν_0, the photons are only slowly absorbed. Suppose that only these (higher energy) photons escape and are absorbed in the outer parts of an HII region. On this basis, do you expect the electron temperature to be higher or lower than in the center of the H II region? Give an explicit argument.

7. Calculate the Rydberg constant for the nuclei of deuterium (^{2}H) and 3-helium (^{3}He), using Eq. (14.19 in 'Tools'), which is:

$$R_M = \frac{R_\infty}{1 + \dfrac{m}{M}}$$

For the electron, the mass is 9.109×10^{-28} g. For D, the nuclear mass is 3.344×10^{-24} g, and for ^{3}He, this is 5.008×10^{-24} g.

8. (a) If the Rydberg constant for ^{4}He is $3.28939118 \times 10^{15}$ Hz, find the separation between the ^{4}He and ^{3}He lines (in km s^{-1}). Given that the linewidths of ^{4}He and ^{3}He are ~ 24 km s^{-1}, and that the number ratio ^{3}He/^{4}He $= 10^{-4}$, sketch the shape of each profile and that of the combined profile.

(b*) For a ^{4}He line T_A of 2 K, frequency resolution 100 kHz, and system noise temperature of 40 K, how long must one integrate using position switching to detect a ^{3}He recombination line?

9. The exact formula for a transition from the ith to the nth level, where $i < k$ is given in Eq. (14.18) in 'Tools', which is:

$$\nu_{ki} = Z^2 R_M \left(\frac{1}{i^2} - \frac{1}{k^2} \right)$$

where $i < k$.
(a) If we set $k = n+1$, show that the approximate Rydberg formula for the transition from the $n + 1^{\text{th}}$ to the nth levels, that is for the $n\alpha$ line, is

$$\nu = \frac{2 Z^2 R_M}{n^3} .$$

(b*) Determine the error for this approximation in the case of $n\alpha$ transitions, for $n = 126, 109, 100$ and 166. If a total analyzing bandwidth of 10 MHz is used to search for these recombination lines, show that the line frequencies calculated using the approximate formula do *not* fall in the spectrometer band.

10. Suppose that the recombination lines from the elements ^{4}He and ^{12}C are emitted without turbulence. The ^{4}He arises from a region of electron temperature, $T_e = 10^4$ K, while the ^{12}C arises from a region with electron temperature $T_e = 100$ K. Eq. (14.20) in 'Tools' is:

$$\Delta V_{\frac{1}{2}} = \sqrt{0.04576 \, T_e + v_t^2}$$

Modify Eq. (14.20), which is valid for H, for these elements using the atomic weights. Assume that the turbulent velocity, v_t, is zero.) The ^{4}He $-^{12}$C separation is 27.39 km s^{-1}. If the intensities are equal, at what level do these lines overlap? Suppose the turbulence of the ^{4}He line is 20 km s^{-1}. Now what is the overlap?

11*. This problem applies the derivation given in Eq. (14.22) to Eq. (14.29) for the recombination of the remaining electron of singly ionized helium. This electron will experience the field of the *doubly ionized* nucleus.
(a) Estimate the line frequencies using the relation in problem 9. Hint: use the Rydberg formula with Z^2 in the numerator.

(b) For measurements of recombination lines of He^+ and He at nearly the same frequency, which line is more intense? For this, one must calculate the dipole moment and A coefficient. The dipole moment as obtained from the correspondence principle is $\mu_{n+1,n} = (1/2)ea_n$, where a_n is the Bohr radius for principal quantum number n (Eq. (12.24) in 'Tools') leads to Eq. (14.22) for hydrogen. Hint: The nuclear charge enters as Z^2 in this relation and in the $An + 1, n$ coefficient.

$$A_{n+1,n} = \frac{64\pi^6 \, m \, e^{10}}{3 \, h^6 \, c^3} \frac{1}{n^5}$$

where the charge on the nucleus is unity. For a larger value, the right side must be multiplied by Z^2. Refer to the previous problem to obtain a_n for an atom with nuclear charge Z. Use the expression for the dipole moment given above, then use this to show that the A coefficient for high $n\alpha$ lines is $A_{n+1,n} = 5.36 \times 10^9 \, n^{-5} Z^2 \, s^{-1}$.

12. Use Eq. (14.32) in 'Tools', which is:

$$T_L = T_{bL} - T_{bc} = T_e e^{-\tau_c} \times (1 - e^{-\tau_L})$$

with $\tau_L \ll 1$ but $\tau_c \gg 1$ to investigate how T_L is affected by a finite continuum optical depth. Suppose you are unaware of the effect of the continuum optical depth; show that the value of T_L is reduced. Use the fact that T_L is proportional to $1/T_e$ to show that the value of T_e obtained from the measurement of T_L and T_c will be larger than the value one would obtain if τ_c is small.

13*. The level populations of hydrogen atoms in an HII region deviate from LTE. We use a specific set of parameters to estimate the size of these quantities. From the Boltzmann relation for $T = 10^4$ K, we find that

$$N(\text{LTE}, 101)/N(\text{LTE}, 100) = 1.00975 .$$

Make use the ratio of the b_n factors, $b_{101}/b_{100} = 1.0011$ for $N_e = 10^3 \, \text{cm}^{-3}$ and $T_e = 10^4$ K (from M. Brocklehurst 1970 Mon. Notices Roy. Astron. Soc. 148, 417), to determine the excitation temperature between the $n = 100$ and $n = 101$ energy levels. Is T_{ex} greater or less than zero? Determine the ratio of b_n factors for $n = 100$ and 101 which allows *superthermal* populations ($T_{ex} > T_k > 0$) by setting $T_e = \infty$.

14. For the $n = 40$ and 41 levels, for $N_e = 10^3 \, \text{cm}^{-3}$ and $T_e = 10^4$ K and $h\nu/k = 4.94$ K, the ratio of $b_{41}/b_{40} = 1.005$. Determine the excitation temperature between these levels.

15*. For $N_e = 10^3 \, \text{cm}^{-3}$ and $T_e = 10^4$ K, for the principal quantum number, $n = 100$, the departure coefficient is $b = 0.9692$ and $(\text{d} \ln b_{100}/\text{d}n)\Delta n = 1.368 \times 10^{-3}$.

Determine β using Eq. (14.42) from 'Tools', which is:

$$\beta = 1 - 20.836 \left(\frac{T_e}{K}\right) \times \left(\frac{\nu}{GHz}\right)^{-1} \times \frac{d \ln b_n}{dn} \Delta n$$

Then calculate $\frac{r}{r^*}$ which is $\frac{T_L}{T_L^*}$ using Eq. (14.52) in 'Tools':

$$\frac{r}{r^*} = \frac{T_L}{T_L^*} = b \times \left(1 - \frac{1}{2} \tau_c \beta\right)$$

where τ_c is the continuum optical depth.

16*. Assume that the carbon 166 α recombination lines, at $\nu = 1.425\,GHz$, are emitted from an isolated region, i.e. without a background source (this is possible since carbon has an ionization potential lower than 13.6 eV). The parameters of this region are $N_C = N_e = 1\,cm^{-3}$, $L = 0.4\,pc$ and $T_e = 100\,K$. Using the LTE relation, what is T_{line} and $T_{continuum}$ if $\Delta\nu = 4.7\,kHz = 1\,km\,s^{-1}$? Now use the appropriate non-LTE coefficients, $b_n = 0.75$, $\beta = -7$, and repeat the calculation.

17. A modified version of the usual equation of radiative transfer is Eq. (14.46) in 'Tools'. This is:

$$-\frac{dI_\nu}{d\tau_\nu} = S_\nu - I_\nu$$

Set the source function, $S \ll I$, to show that $I = I_0 \, e^{\kappa_\nu \beta b \, L}$. This is the situation in which there is an intense background source with $T_{BG} \gg T_e$. Then repeat Problem 19 for $T_{BG} = 2500\,K$.

18. There are a few neutral clouds along the line of sight to the supernova remnant Cassiopeia A. Assume that these are the only relevant carbon recombination line sources. These clouds are known to have H_2 densities of $\sim 4 \times 10^3\,cm^{-3}$, column densities of $\sim 4 \times 10^{21}\,cm^{-2}$, diameters of 0.3 pc. If we assume that *all* of the carbon is ionized, we have $C^+/H = 3 \times 10^{-4}$. At wavelengths of more than a few meters, the carbon lines are in absorption. Assume that the line formation is hydrogen-like. For the C166α line, we estimate that the peak line temperature is 3 K, and the Doppler FWHP is $\Delta V_{1/2} = 3.5\,km\,s^{-1}$ (see Kantharia et al. (1998 Ap. J. 506, 758) for a model and references).
(a) Show that for $n > 300$ collisions dominate radiative decay, so that the populations are thermalized, but that the populations are dominated by radiative decay for $n < 150$.
(b) An observer claims that "Since the C166α line is in *emission*, the excitation temperature must be larger than the background temperature, or negative". Do you agree or disagree? Cite equations to justify your decision.

(c) If $h\nu \ll kT$ and LTE conditions hold, show that a reformulation of Eq. (14.28) in 'Tools', which is:

$$T_{\mathrm{L}} = 1.92 \times 10^3 \times (T_e)^{-3/2} \times (\frac{\mathrm{EM}}{\Delta \nu})$$

where EM, the Emission Measure, is in units of cm^{-6} pc, and $\Delta \nu$ is in units of kHz. This gives the following relation between EM and line intensity:

$$T_{\mathrm{L}} = \left(\frac{576}{T_e^{3/2}}\right) \times \left(\frac{\mathrm{EM}}{\nu_0 \, \Delta \mathrm{V}_{1/2}}\right) ,$$

where $\Delta \mathrm{V}_{1/2}$ is in $\mathrm{km\,s}^{-1}$, ν_0 in GHz, EM in cm^{-6} pc and all temperatures in Kelvin.

(d)* Estimate the maximum brightness temperature of Cas A at 1.425 GHz. Assuming that the level populations are *not* inverted, the excitation temperature, T_{ex}, of the C166α transition is >0. Given the background continuum temperature of Cassiopeia A, estimate a lower limit for T_{ex} from the continuum brightness temperature of Cassiopeia A. Next, use the cloud parameters in part (a) to determine the emission measure of C^+. Finally, make use of the expression in part (c) to determine the integrated C166α line intensity. From a comparison with the observed result, is it more reasonable to assume that $T_{\mathrm{ex}} > 0$ or that population inversion is more likely? In this case, population inversion will give rise to line masering effects.

Chapter 15
Overview of Molecular Basics

1. (a) For $T = 273$ K and pressure 1 atmosphere, that is 10^6 dyne cm^{-2} (760 mm of Hg), find the density, n, of an ideal gas in cm^{-3}. Repeat for conditions in a molecular cloud, that is $T = 10$ K, pressure 10^{-12} mm of Hg.
(b) For both sets of conditions, find the mean free path, λ, which is defined as $1/(\sigma\, n)$, and the mean time between collisions, τ, which is $1/(\sigma\, n\, v)$, where v is the average velocity. In both cases, take $\sigma = 10^{-16}$ cm^{-3}. For the laboratory, take the average velocity to be 300 m s^{-1}; for the molecular cloud, take the average velocity of H_2 as 0.2 km s^{-1}.
(c) Suppose that the population of the upper level of a molecule decays in 10^5 s. How many collisions in both cases occur before a decay?
(d) For extinction we define the penetration depth, λ_v, in analogy with the mean free path. When $\lambda_v = 1$ the light from a background star is reduced by a factor 0.3678. For a density of atoms n, λ_v, in cm, is $2 \times 10^{21}/n$. Calculate the value of λ_v for a molecular cloud and for standard laboratory conditions. The parameters for both are given in part **(a)** of this problem.

2. (a) The result of problem 2(c) of chapter 13 gives $T_k = 21.2\,(m/m_H)\,(\Delta V_t)^2$ where ΔV_t is the FWHP thermal width, i.e. there is no turbulence and the gas has a Maxwell–Boltzmann distribution. Apply this formula to the CO molecule (mass $28\,m_H$) for a gas of temperature T. What is ΔV_t for $T = 10$ K, $T = 100$ K, $T = 200$ K?
(b) The observed linewidth is 3 km s^{-1} in a dark cloud for which $T = 10$ K. What is the turbulent velocity width in such a cloud if the relation between the observed FWHP linewidth, $\Delta V_{1/2}$, the thermal linewidth, ΔV_t and the turbulent linewidth ΔV_{turb} is

$$\Delta V_{1/2}^2 = \Delta V_t^2 + \Delta V_{turb}^2\,?$$

© Springer International Publishing AG, part of Springer Nature 2018
T. L. Wilson, S. Hüttemeister, *Tools of Radio Astronomy – Problems
and Solutions*, Astronomy and Astrophysics Library,
https://doi.org/10.1007/978-3-319-90820-5_15

3. The following expression is appropriate for the spontaneous decay between two rotational levels, (u, l) of a linear molecule: $A_{ul} = 1.165 \times 10^{-11} \mu_0^2 \nu^3 (J+1)/(2J+3)$ where ν is in GHz, μ_0 is in Debyes and J is the lower level in the transition from $J+1 \to J$. Use this to estimate the Einstein A coefficient for a system with a dipole moment of 0.1 Debye for a transition from the $J=1$ level to the $J=0$ level at 115.271 GHz.

4. To determine whether a given level is populated, one frequently makes use of the concept of the "critical density", n^*, defined as:

$$A_{ul} = n^* \langle \sigma v \rangle .$$

where u is the quantum number of the upper rotational level, and l is that for the lower level. If we take $\langle \sigma v \rangle$ to be $10^{-10} \, \text{cm}^3 \, \text{s}^{-1}$, determine n^* from the following A_{ul} coefficients

$$CS : A_{10} = 1.8 \times 10^{-6} \, \text{s}^{-1}$$

$$CS : A_{21} = 2.2 \times 10^{-5} \, \text{s}^{-1}$$

$$CO : A_{10} = 7.4 \times 10^{-8} \, \text{s}^{-1}.$$

5. Suppose the effective radius $r_e = 1.1 \times 10^{-8}$ cm and the reduced mass, m_r, of a perfectly rigid molecule is 10 atomic mass units, AMU (an AMU is 1/16 of the mass of a 16-oxygen atom; 1 AMU= 1.660×10^{-24} g), where $\Theta = m_r r_e^2$.
(a) Calculate the lowest four rotational frequencies and energies of the levels above the ground state. One needs a simplified version of Eq. 15.11 to 15.13 from 'Tools' for the rotational constant is

$$B_e = \frac{\hbar}{4\pi \, \Theta_e} \tag{15.1}$$

The energy of level J is:

$$E_{rot} = W(J) = \frac{\hbar^2}{2 \, \Theta_e} J(J+1) - hD\,[J(J+1)]^2 . \tag{15.2}$$

and the frequency is the difference between the energy of level $J+1$ and J divided by the Planck constant:

$$\nu(J) = \frac{1}{h} [W(J+1) - W(J)] = 2 B_e \left[(J+1) - 4D(J+1)^3 \right] \tag{15.3}$$

(b) Repeat if the reduced mass is (2/3) AMU with a separation of 0.75×10^{-8} cm; this is appropriate for the HD molecule. The HD molecule has a dipole moment $\mu_0 = 10^{-4}$ Debye, caused by the fact that the center of mass is not coincident with

Table 15.1 Parameters of the more commonly observed carbon monoxide lines (problem 6)

Chemical[a] formula	Molecule name	Transition	ν/GHz[b]	E_u/K[c]	A_{ij}/s^{-1}[d]
C^{18}O	Carbon monoxide	$J = 1 - 0$	109.782182	5.3	6.5×10^{-8}
^{13}CO	Carbon monoxide	$J = 1 - 0$	110.201370	5.3	6.5×10^{-8}
CO	Carbon monoxide	$J = 1 - 0$	115.271203	5.5	7.4×10^{-8}
C^{18}O	Carbon monoxide	$J = 2 - 1$	219.560319	15.9	6.2×10^{-7}
^{13}CO	Carbon monoxide	$J = 2 - 1$	220.398714	15.9	6.2×10^{-7}
CO	Carbon monoxide	$J = 2 - 1$	230.538001	16.6	7.1×10^{-7}

[a] If isotope not explicitly given, this is the most abundant variety, i.e., ^{12}C is C, ^{16}O is O, ^{14}N is N
[b] From Lovas (1992, J. Chem. Phys. Ref. Data 21, 18)
[c] Energy above the ground state in Kelvins
[d] Spontaneous transition rate, i.e., the Einstein A coefficient

the center of charge. Take the expression for $A(ul)$ from Problem 3 and apply to the $J = 1 - 0$ and $J = 2 - 1$ transitions of the HD molecule.
(c) Find the "critical density", $n^* \approx 10^{10} A(ul)$.

6. The ^{12}C^{16}O molecule has $B_e = 57.6360$ GHz and $D_e = 0.185$ MHz. Calculate the energies for the $J = 1, 2, 3, 4, 5$ levels and line frequencies for the $J = 1 - 0, 2 - 1, 3 - 2, 4 - 3$ and $5 - 4$ transitions. Use the expression energy $E(J)/h \approx B_e J(J + 1) - D_e J^2(J + 1)^2$ for the energy calculation. Check the results against the relevant parts of Table 16.1 in 'Tools', given here as Table 15.1.

7. Apply for $J = 0, 1$ the analysis in problem 6 to the linear molecule HC$_{11}$N, which has $B_e = 169.06295$ MHz and $D_e = 0.24$ Hz. Estimate J for a transition near 20 GHz. What is the error if one neglects the distortion term?

8. In the following, we neglect the distortion term D_e and assume that the population is in LTE. The population in a given J level for a linear molecule is given by Eq. (15.33):

$$n(J)/n(\text{total}) = (2J + 1)e^{B_0 J(J+1)/kT}/Z$$

where Z, the partition function, does not depend on J. Differentiate $n(J)$ with respect to J to find the state which has the largest population for a fixed value of temperature, T. Calculate this for CO if $T = 10$ K and $T = 100$ K. Repeat for CS ($B_0 = 24.584$ GHz) and HC$_{11}$N, for $T = 10$ K.

9. Extend Eq. (15.33 in 'Tools'), which is:

$$N(J)/N(\text{total}) = \frac{(2J + 1)}{Z} \exp\left[-\frac{h B_e J(J + 1)}{kT} \right]$$

to include the optical depth relation Eq. (15.26), which is:

$$N_l = 1.95 \times 10^3 \frac{g_l \nu^2}{g_u A_{ul}} \int T_B \, dV$$

to obtain an estimate of which J level has the largest optical depth, τ, in the case of emission for a linear molecule.

(a) Show that when the expression for the A coefficient for a linear molecule is inserted into Eq. (15.26 of 'Tools'), we have

$$N_1 = \frac{1.67 \times 10^{14}}{\mu_0^2 \, v[\text{GHz}]} \times \frac{2J+1}{J+1} \, T_{\text{ex}} \, \tau \Delta v \,,$$

where μ is in Debyes and v is in km s^{-1}.

(b) Use the above expression to estimate whether the J for the maximum $T_{\text{MB}} = T_{\text{ex}} \tau$ is larger or smaller than the J obtained in Problem 8.

10. Find the ratio of the intensities of the $J = 2 - 1$ to $J = 1 - 0$ transitions for a linear molecule if the excitation temperature of the system, T, is very large compared to the energy of the $J = 2$ level above the ground state, and both lines are optically thin. What is the ratio if both are optically thick? Use the last equation in the statement of Problem 9 of this Chapter.

11. The ammonia molecule, NH_3, is an oblate symmetric top. For ammonia, $B = 298\,\text{GHz}$, $C = 189\,\text{GHz}$. If $T \gg B, C$, the value of Z, the partition function, with C and B in GHz, is $Z = 168.7\sqrt{(T^3)/(B^2\,A)}$.

(a) Evaluate Z for NH_3 for $T = 50\,\text{K}$, $100\,\text{K}$, $200\,\text{K}$, $300\,\text{K}$. For this approximation to be valid, what is a lower limit to the value of T?

(b) The (3,3) levels are $120\,\text{K}$ above ground. Use the partition function and

$$n(J)/n(\text{total}) = (2J+1)e^{120/T}/Z$$

to calculate the ratio of the total population to that in the (3,3) levels.

(c) If only metastable ($J = K$) levels are populated, use the definition of Z as a sum over all populated states, and

$$n(J)/n(\text{total}) = (2J+1)e^{(BJ(J+1)+K^2(C-B))/kT}/Z$$

and the B and C values for NH_3 to obtain the ratio between the population of the (3,3) levels and all metastable levels.

12. The selection rules for dipole transitions of the doubly deuterated isotopomer D_2CO differ from that of H_2CO since D_2CO has two Bosons, so the symmetry of the total wavefunction must be symmetric. Determine these rules following the procedure in Sect. 15.6.2.

Chapter 16
Molecules in Interstellar Space

1. For CH_3CN, CH_3C_2H and NH_3 there can be no radiative transitions between different K ladders. The populations can however be exchanged via collisions. For ammonia, there must be $J > K$. There is a rapid decay of populations with quantum numbers from $(J + N + 1, K)$ to $(J + N, K)$, where N is ≥ 1. Use the result in problem 11, chapter 15 to show that rotational transitions of NH_3 fall in the frequency range $\geq 500\,\text{GHz}$. Estimate the Einstein A coefficients for the $J = 1 - 0$ and $J = 2 - 1$ transitions using $\mu = 1.34$ Debye. Compare these values to those for the inversion transitions listed in Table 16.1, which are $\sim 10^{-7}\,\text{s}^{-1}$.

2.(a) Calculate the excitation temperature, T_{ex}, between two energy levels which have the same statistical weights, that is $g_u = g_l$, so that the Boltzmann equation is $n_u/n_l = e^{-h\nu/kT_{\text{ex}}}$ with $h\nu/k = 1.14\,\text{K}$ (see also problem 1 of chapter 12 where these parameters are defined). The values of n_u/n_l are 0.5, 0.6, 0.7, 0.8, 0.9, 1.0, 1.1, 1.2, 1.3, 1.4, 1.5.

(b) Use the relation between optical depth and column density from Problem 9(a) of chapter 15 to calculate the optical depth, τ, for the $J = 1 - 0$ line which has a FWHP of $10\,\text{km s}^{-1}$, $T_{\text{ex}} = -100\,\text{K}$, $\mu_0 = 3.6$ Debye, and $\nu = 9.0\,\text{GHz}$.

(c) Substitute this value of τ into the relation $T_{\text{MB}} = (T_{\text{ex}} - T_{\text{BG}})(1 - e^{-\tau})$. If $|T_{\text{BG}}| \ll |T_{\text{ex}}|$ and $|\tau| \ll 1$, show that T_{MB} gives an accurate estimate of the column density in the lower level, N_0. Aside from questions of English usage, would you agree with the statement "Optically thin masers do not mase"?

(d) Evaluate the other extreme case, $T_{\text{BG}} \gg T_{\text{ex}}$, to show that the background radiation is amplified.

© Springer International Publishing AG, part of Springer Nature 2018
T. L. Wilson, S. Hüttemeister, *Tools of Radio Astronomy – Problems and Solutions*, Astronomy and Astrophysics Library, https://doi.org/10.1007/978-3-319-90820-5_16

3. Use the large velocity gradient (LVG) relation for a two-level system Eq. (16.39) in 'Tools', which is:

$$\frac{T}{T_0} = \frac{T_k/T_0}{1 + T_k/T_0 \ln\left[1 + \dfrac{A_{ji}}{3C_{ji}\,\tau_{ij}}\left(1 - \exp\left(-3\,\tau_{ij}\right)\right)\right]}$$

to estimate the line temperature when $T_K \gg T_0$, $A \gg C$. In addition, $(A_{ji})/(3C_{ji}\,\tau_{ij}) \ll 1$. This is a hot, subthermally excited transition.

4. Repeat the above exercise for the case in which $A \ll C$, but with all other parameters unchanged. This is the case of a hot, thermalized gas. Compare these results with those of Problem 3.

5*. In circumstellar envelopes, one assumes that spherical symmetry holds, and that density $n(r) = n_0 r^{-2}$. In addition, $r = (z^2 + p^2)^{1/2}$, where p is the projected distance and z the line-of-sight distance, and a constant velocity of expansion.
(a) Show that

$$\delta z = \Delta V/(dv_{\parallel}/dz) = p\frac{\Delta V}{V}(1 - (v_{\parallel}/V)^2)^{-3/2} .$$

(b) Take the abundance of a species to be a constant fraction of the abundance of H_2. Show that the optical depth for a given species at point p and for a given $v_{\parallel}$ is

$$\tau(p, v_{\parallel}) = \frac{\mu^2 f n_0 (J+1)}{1.67 \times 10^{14} \, T_{ex}} \frac{p}{V(2J+1)}(1 - (v_{\parallel}/V)^2)^{-1/2} .$$

(c) Take the beam to be much larger than the source. Then show that

$$T = 2\pi \int_0^{p_{max}} T_0(1 - e^{-\tau})\, p \, dp$$

(d) Assume that the line is optically thin. Show that the line profile is flat–topped. Then assume that the line is optically thick. Show that the profile has a parabolic shape.

6*. Bipolar outflows are common in pre-main sequence sources. This is a very elementary analysis of molecular line emission from well-defined bipolar outflows.
(a) Approximate the outflow as a cylinder of length l, width w, with constant density n, inclined at an angle i to the line of sight. Show that the functional description of the mass of the outflowing material is $(1/4)n(H_2)\pi\, l\, w^2$.
(b) If the observed velocity of the outflow is v_0, show that the age of the outflow is

$$\text{age} = l_0/(v_0 \tan i) .$$

(c) Show that the total kinetic energy in the outflow is $(1/2)\,Mv_0^2\sin i/\cos^2 i$.

(d) If we define the mechanical luminosity L as $\dot{E} = (2\times$ kinetic energy in the outflow)/age, show that $L = M(M_\odot)\,v^3/(\sin i\cos^3 i)$, where M is the mass of the outflow.

7*. For linear molecules, in principle one can determine both the kinetic temperature and the H_2 density if one can measure the "turn over" in the distribution of column densities from different transitions. One example is given by measurements of the CO molecule in Orion KL.

(a) You need the results in problems 3, 4 and 9 of chapter 15 to solve this problem. Estimate the wavelengths, frequencies and Einstein A coefficients for the $J = 30 - 29$, $J = 16 - 15$ and $J = 6 - 5$ transitions, if CO is a rigid rotor molecule. Compare these to the value for the $J = 2 - 1$ transition. If the lines are optically thin, and $\langle\sigma v\rangle = 10^{-10}$ cm^3 s^{-1}, what are the critical densities?

(b) Determine the energies of the $J = 30$, $J = 16$ and $J = 6$ levels above the ground state. If the kinetic temperature of this outflow region is $\sim$2000 K, find the ratio of populations of the $J = 30$ to $J = 6$ levels, assuming LTE conditions. If the H_2 density, n, is $\sim10^6$ cm^{-3}, set A equal to the collision rate, $C = n\langle\sigma v\rangle$, to determine which of the transitions is sub-thermally excited, i.e. $A \gg C$.

8. For interstellar grains, one can assume a size of 0.3 μm and an abundance of 1 grain for every 10^{13} hydrogen atoms. Then show that the quantity $\sigma_g n_g$ equals $10^{-22}n_H$ cm^{-1}. The mean free path, λ, is equal to $1/\sigma_g n_g$. If the mean time between collisions, $t_{gas-grain} = \lambda/V$, where the expression for V is taken to be $\Delta V_{1/2}$. Show that this leads to Eq. (16.53) namely,

$$t_{gas-grain} = \frac{1.2\times 10^{10}}{n_{H_2}} \qquad (16.1)$$

9. From Eq. (16.50 in 'Tools'), the free-fall time in years for a cloud under the influence of self gravity. This is $t_{ff} = 5\times10^7/\sqrt{n(H_2)}$, where $n(H_2)$ is the molecular hydrogen density in cm^{-3}. From this result and the result in the previous problem, find the density at which the free-fall time equals the average time for a molecule to strike a grain.

10. A typical giant molecular cloud (GMC) is thought to have a diameter of 30 pc, and total mass of $10^6 M_\odot$. Assume that GMC's have no small scale structure.

(a) Develop a general formula relating the H_2 density to the mass and radius of a uniform spherical cloud. Because the He/H number ratio is 0.1, the average molecular mass is 4.54×10^{-24} g.

(b) What is the density of the GMC? Find the column density of H_2 in this cloud. If the visual extinction is related to the column density of H_2 by $1^m = 10^{21}$ cm^{-2}, what is the extinction through the GMC?

(c) What is the FWHP width of a line if the cloud is in virial equilibrium? Use the simplest condition for virial equilibrium, as given in (Eq. (13.72) in 'Tools'), which is:

$$\frac{M}{M_\odot} = 250 \left(\frac{\Delta V_{1/2}}{\text{km s}^{-1}}\right)^2 \left(\frac{R}{\text{pc}}\right)$$

(d) If the mass of the ISM in the galaxy from 2 kpc to 8.5 kpc is $3 \times 10^9 \, M_\odot$, and if there are ~ 100 GMCs as described in part (a) in this part galaxy, how much of the total mass of the interstellar medium is in GMCs? If the thickness of the galaxy is 200 pc, how much of the volume is contained in GMCs?

(e) What is the H_2 column density through a GMC? If one visual magnitude is equivalent to a column density of $10^{21} \, \text{cm}^{-2}$ of H_2, what is A_v of a GMC?

11. (a) A well-established ion exchange reaction in diffuse molecular clouds is

$$H_2 + D^+ = HD + H^+ + \Delta E, \tag{16.2}$$

where the zero point energy difference between H_2 and HD is $\Delta E / k = 500$ K. It is thought that reaction (16.2) reaches equilibrium. Then one can relate the initial and final products by the Boltzmann relation

$$\frac{HD}{H_2} = \frac{D^+}{H^+} e^{\Delta E / kT} \, .$$

If the relevant temperature, T, is $T_k = 100$ K, what is the overabundance of HD?

(b) A similar reaction to that given in part (a) occurs for isotopes of carbon monoxide (CO) if the carbon ion, C^+, is present in the outer parts of molecular clouds:

$$^{12}CO + {}^{13}C^+ = {}^{13}CO + {}^{12}C^+ + \Delta E \, .$$

In this case, $\Delta E / k = 35$ K. Repeat the steps in part (a) for the case of CO and ^{13}CO.

Chapter 17
Solutions for Chapter 1: Radio Astronomical Fundamentals

1. The equation given in the statement of this problem determines the cutoff frequency of a plasma. For the Interstellar Medium (ISM), that is, $\nu_p = 8.97 \times (0.03)^{0.5} = 1.6\,\text{kHz}$, while the ionospheric cutoff is $\nu_p = 8.97(10^5)^{0.5} = 2.8\,\text{MHz}$, from the discussion on page 4 of 'Tools'. Since the plasma cutoff frequency in the ISM is much lower than the cutoff frequency in the Earth's ionosphere, there may be astronomical phenomena which can be observed only from above the ionosphere. Thus such measurements must be made from satellites.

2. (a) From the result of problem 1, this radiation must arise in the ionosphere.
(b)* Assuming that this is black body radiation which falls on the antenna from one hemisphere (solid angle $\Omega = 2\pi$ steradians), we use the Rayleigh–Jeans law to relate the brightness temperature to the power. The antenna can receive only one polarization, so the relation for power, P, is $P = S_\nu \times$ Area $\times$ Bandwidth $\times$ Solid Angle and $S_\nu = 2kT/\lambda^2 \times$ Solid Angle: $P = (2kT/\lambda^2)AB(1/2)2\pi$. For these values, the resulting power is 1.5×10^{-23} W.

3. The peak flux density is
$S = 1500\,\text{W}/[2\pi(5 \times 10^5\,\text{m})^2 \times (10^9\,\text{Hz})] = 9.6 \times 10^{-19}\,\text{W\,m}^{-2}\,\text{Hz}^{-1}$
$= 9.6 \times 10^7$ Jy. The average power is 3% of this value or 2.9×10^6 Jy.

4. Inserting the values give $S = 600$ W$/[4\pi(10\ \text{m})^2(10^6\,\text{Hz})] = 4.8 \times 10^{-7}\,\text{Wm}^{-2}\,\text{Hz}^{-1} = 4.8 \times 10^{19}$ Jy.

5. (a) In the microwave band, the flux is $S = 1000\text{W}/[4\pi(3.84 \times 10^8\ \text{m})^2 \times (3 \times 10^8\,\text{Hz})] = 1.8 \times 10^{-24}\,\text{W\,m}^{-2}\,\text{Hz}^{-1} = 1.8 \times 10^2$ Jy. In the optical range, the bandwidth is much larger (5×10^{14} Hz), resulting in a far smaller value, $S = 5 \times 10^{-4}$ Jy.
(b) In the microwave range, the average photon energy is $E_r(\text{photon}) = (6.62 \times 10^{-27}\ \text{erg s}) \times (2.8 \times 10^9\,\text{Hz}) = 1.85 \times 10^{-17}\,\text{erg} = 1.85 \times 10^{-24}$ J. In the optical, it is $E_o(\text{photon}) = 3.6 \times 10^{-12}\ \text{erg} = 3.6 \times 10^{-19}$ J. The total number of photons emitted per second is $N_r = 5.49 \times 10^{27}$ (radio) and $N_o = 3.0 \times 10^{21}$ (optical).

© Springer International Publishing AG, part of Springer Nature 2018
T. L. Wilson, S. Hüttemeister, *Tools of Radio Astronomy – Problems and Solutions*, Astronomy and Astrophysics Library,
https://doi.org/10.1007/978-3-319-90820-5_17

The photon flux can be calculated using $N(\text{photon})/A$, where A is the area of the sphere with the radius being the distance to the Moon. The photon fluxes are $3.0 \times 10^8\,\text{m}^{-2}\,\text{s}^{-1}$ (radio) and $1620\,\text{m}^{-2}\text{s}^{-1}$ (optical), respectively.

6. At 1 km, $S = 10^{-9}\,\text{W}\,\text{m}^{-2}\,(3/1000)^2/(10^8\,\text{Hz})] = 9 \times 10^{-23}\,\text{W}\,\text{m}^{-2}\,\text{Hz}^{-1}$ The result is 9×10^3 Jy. If there is a line of sight to the telescope, there will be interference at a level of at least the sensitivity limit of the telescope up to a distance of 3000 km, if there is no atmospheric absorption. The actual distance will be smaller if the radiation follows a straight line path. For this distance, D, if the height is small compared to the radius of the Earth, R_e, we have

$$D = \sqrt{2\,h\,R_e}$$

For a height, h, of 1 km, the range of the interference would be limited to 80 km.

7. We have $\nu_{\max} = 3.4 \times 10^{14}\,\text{Hz} = 3.4 \times 10^5\,\text{GHz} = 58.789\,T$, thus $T = 5780\,\text{K}$. The number of photons per second is n. In the (simple but rather unrealistic) monoenergetic case, we use $n = L_\odot/h\nu = 1.7 \times 10^{45}$ photons emitted per second. At $1\,\text{AU} = 1.5 \times 10^{13}$ cm, this number is reduced by a factor of $4\pi(1.5 \times 10^{13})^2 = 2.8 \times 10^{27}\,\text{cm}^2$, so the number received on earth is $6.0 \times 10^{17}\,\text{cm}^{-2}\,\text{s}^{-1}$. Multiplying by the energy of each photon, 2.25×10^{-12} erg, we have $1.4 \times 10^6\,\text{erg}\,\text{cm}^{-2}\,\text{s}^{-1} = 0.14\,\text{W}\,\text{cm}^{-2} = 1.4\,\text{kW}\,\text{m}^{-2} = 1.4\,\text{mW}\ \text{cm}^{-2}$. So, good agreement. Problems number 8 and 9 were shifted to later chapters.

10. From the Rayleigh-Jeans relation in MKS units, we have from Eq. (1.33):

$$S_\nu = 2k\,T_B\,\theta^2/\lambda^2\,\Delta\Omega$$

where $\delta\Omega$ is the solid angle in steradians. The value of Boltzmann's constant, k, is 1.38×10^{-23} The area under a Gaussian in one dimension is 1.06 times the product of the height and width. For two dimensions, this is 1.13. For a normalized Gaussian in two dimensions, we take the angle in steradians to be $1.13 \times \theta^2$, where θ is in radians. Then we have $\theta(\text{arcmin}) = \frac{\pi}{180} \times \frac{1}{60}\theta(\text{rad}) = $ and for wavelength, $\lambda(\text{m}) = \frac{1}{60}\lambda(\text{cm})$ combining these, we have Eq. (1.34):

$$S_\nu = 2.65\,T_B \left(\frac{\theta}{\text{arcmin}}\right)^2 (\frac{\lambda}{\text{cm}})^{-2}$$

Using the standard relations of angles, mm and cm, we used $2.65 \times 3600/100$ to obtain Eq. (1.44):

$$\left(\frac{S_\nu}{\text{mJy}}\right) = 73.6\,T_B \left(\frac{\theta}{\text{arcsec}}\right)^2 \left(\frac{\lambda}{\text{mm}}\right)^{-2}$$

11. Use the result of problem 10 or Eq. (1.34):

$$S_\nu(\text{Jy}) = 2.65 \times 10^5 \times (24 \times 60)^2/(14.6 \times 10^2)^2 = 2.58 \times 10^5\,\text{Jy}$$

For the power, P, we multiply S_ν by the collecting Area, A and bandwidth, B. The result is $P = 2.58 \times 10^5 \times 10^3\text{Hz} \times 800\text{m}^2$. Then the power is: $P = 2.1 \times 10^{-13}\,\text{W}$.

12. Use the result of problem 10 or Eq. (1.34) in 'Tools'. The wavelength is 15 meters, and the angle is 24°, so: $S_\nu(\text{Jy}) = 2.65 \times 7 \times 10^4 \times (24 \times 60)^2/(15.0 \times 10^2)^2 = 1.7 \times 10^5\,\text{Jy}$. The source shape is a ring, but considering this much larger telescope beam, can be approximated as a Gaussian. Then the relation holds, and $T_{\text{source}} \times 4^2 = 7 \times 10^4 \times (24 \times 60)^2 = 9.1 \times 10^9\,\text{K}$. As will be shown in chapter 10 and figure 10.1, where source flux densities are plotted as a function of frequency. This plot shows that the Cas A emission is non-thermal.

The reader will have noted that we have used a convenient property of Gaussian functions to obtain the actual source temperature. The proof of this is a long drawn out process, which we give in problem 6 of chapter 8 of this volume. One may wonder why this is only carried out in chapter 8 where the use of telescopes is presented. The most important result is that the observed antenna temperature, T_A, is the convolution (see chapter 4 of this volume) of the telescope power pattern, P, with the actual source temperature, T_B. This is presented in problem 1 of chapter 7. Not all sources have a Gaussian shape, but can be approximated by the sum of Gaussians. Finally, the 'actual' temperature of a source is an approximation which depends on the angular resolution; the higher the angular resolution, the more detail of a source is revealed. As will be shown, this detail is limited by the ratio of the flux density in a beam element to the receiver noise.

13*. For the version of the Planck formula in wavelength units, we have from Eq. (1.22):

$$B_\lambda(T) = \frac{2hc^2}{\lambda^5}\frac{1}{e^{hc/k\lambda T} - 1}$$

where $y = 1.44 \cdot 10^4/\lambda(\mu\text{m})\,T$. The expression $\frac{2hc^2}{\lambda^5}$ determines the coefficient. In MKS units, this is:
$1.19 \times 10^8/\lambda(\mu)^5$. In W cm^{-2} Steradians^{-1} μ^{-1}, the value, 3.7×10^{-7}, is a factor of about 3 larger than that value extracted from the plot in Leighton, Appendix B, p. 725.

Changing to units of ergs cm^{-2} μm^{-5} Hz^{-1}, this has a value of 1.19×10^{15} erg cm$^{-2}\mu$m^{-1}.

In CGS units, that is, ergs s^{-1} cm^{-2} Hz^{-1} steradian^{-1}, the coefficient is 7.36×10^{-21} The conversion is $d\nu/\nu = \Delta V/c$, from the Doppler relation. If the units are GHz and km s^{-1},

$$d\nu = (10^9\nu(\text{GHz})) \times dV(\text{km s}^{-1})/(3 \times 10^5) = 3.3 \times 10^3 dV(\text{km s}^{-1})$$

the relation is

$$\int B_\nu dV = 4.9 \times 10^{-17} \times (\nu(GHz))^3 \frac{1}{e^y - 1} \Delta V(km\, s^{-1})$$

with y= $h\nu/kT = 4.84 \times 10^{-2}\, \nu(GHz)/T$.

14. The expression for the Planck formula is: $B_\nu(T) = \frac{2h\nu^3}{c^2} \frac{1}{e^{h\nu/kT}-1}$. In terms of the frequency in GHz, the coefficient is:

$$\frac{2h\nu^3}{c^2} = 2 \times 6.62 \times 10^{-34}\, 10^{27}\, \nu(GHz)^3/(3 \times 10^8)^2 = 1.47 \times 10^{-23}$$

the value of 'y' is given in the previous problem. So,

$$\left(\frac{B_\nu(T)}{Jy\, sterad}\right) = 1.47 \times 10^3 \times \nu(GHz)^3 \times \frac{1}{e^{4.8 \times 10^{-2}\, \nu(GHz)/T} - 1}$$

In units of $W\, m^{-2}\, km\, s^{-1}$ and ν in GHz, the term $\frac{2h\nu^3}{c^2}$ is:

$$2 \times 6.62 \times 10^{-34} \times 10^{27}\nu(GHz)^3/(3 \times (10^8)^2) = 7.36 \times 10^{-24}\, \nu(GHz)^3$$

In the Rayleigh-Jeans limit, this becomes,

$$\left(\frac{B_\nu(T)}{Jy\, sterad}\right) = 7.36 \times 10^2 \times \nu(GHz)^3 \times \frac{1}{e^{4.8 \times 10^{-2}\, \nu(GHz)/T} - 1}$$

In other units,

$$\left(\frac{B_\nu(T)}{ergs\, s^{-1}\, cm^{-2}\, sterad^{-1}}\right) = 4.81 \cdot 10^{-17}\, [\nu(GHz)]^4\, \Delta V(km\, s^{-1}) \frac{1}{e^x - 1}$$

where the term x is given as before. In the Rayleigh-Jeans limit, this is:

$$\left(\frac{B_\nu(T)}{ergs\, s^{-1}\, cm^{-2}\, sterad^{-1}}\right) = 10^{-15}\, (\nu(GHz))^3\, T\, \Delta V(kms^{-1})$$

15.[*] Assume that the emission from Jupiter has a Gaussian shape. Then we can use Eq. (1.33): $S_\nu = 2.65\, T_B \left(\frac{\theta}{arcmin}\right)^2 \times \left(\frac{\lambda}{cm}\right)^{-2}$ with a size of θ=0.67 arcmin, and 1.4 GHz=λ=21 cm, so that $S_\nu(Jy) = 2.65 \times 150 \times (0.67)^2/(21)^2 = 0.40\, Jy$. For ν=115 GHz, λ=0.3 cm, so scaling from the last result, $S_\nu(Jy) = 1983\, Jy$. This illustrates the behavior of black bodies, for which the Flux Density rises with increasing frequency squared. In actual fact, in addition, Jupiter exhibits non-thermal radiation at lower frequencies.

The flux density relation for Orion A, the flux density of the core at $\nu=4.8$ GHz or $\lambda=6.2$ cm, requires a correction for the 2.6' beam size. This requires the use of the relation in problems 10 and 12. Use of this gives an actual peak source temperature of 687 K. Then: $S_\nu(Jy) = 2.65 \times 687 \times 2.5^2/6.2^2 = 300$ Jy. (For a comparison, see figure 10.1 in chapter 10 or Fig. 10.1 in 'Tools').

16. From the definition of a solid angle, Ω, this is

$$\Omega = 2\pi \int \sin\theta d\theta = 2\pi \left(1 - \cos\theta\right)$$

For a small angle, $\Omega = \pi\theta^2$, where θ is in radians. For $R_J=71,492$ km, and 1 pc=3 $\times 10^{13}$ km, this source subtends a solid angle of 3.8×10^{-17} steradians. From the result of problem 13, the intensity in Jy/steradians is:

$$I_\nu(Jy) = 1.47 \times 10^3 \times \nu(GHz)^3 \; \frac{1}{e^x - 1}$$

where $x = 4.8 \times 10^{-2} \, \nu(GHz)/T$. Then the product of Planck intensity and angular size is as given. Inserting the values:

$$S_\nu = 2.6 \times 10^{-14}(345)^3 \times \frac{1}{(e^{0.0066} - 1)}$$

The result is: $2.6 \times 10^{-14} \times 150.2 \times (345)^3 \times (20/30)^2 = 71 \, \mu Jy$

17. Following the approach used in problem 16, and the values in the table following the Preface, the solid angle is 1.69×10^{-15}. Combing this with the result of problem 13, we have:

$$S_\nu = 1.47 \times 10^3 \times \nu(GHz)^3 \times \left(\frac{1}{e^x - 1}\right) \times \left(1.69 \times 10^{-15}\right) \times \left(\frac{R_\odot}{D(pc)}\right)^2$$

The coefficient is 2.48×10^{-12} and $x = 0.0029$ Assuming that 0.8 mm=345 GHz, this is:

$$S_\nu = 2.48 \times 10^{-12} (345)^3 \times (1/20)^2 \frac{1}{(e^{0.0029} - 1)} = 80 \, \mu Jy$$

The distance must be reduced by the square root of the ratio of 88 µJy to 1 milli Jy. This gives a distance of 3.4 pc.

18. Use figure 1.1 (Fig. 1.6 in 'Tools'; see statement of this problem) to get a value for 10^{13} Hz $= 10^4$ GHz. Make use of Eq. (1.25), which is:

$\left(\frac{\nu_{max}}{GHz}\right) = 58.789 \left(\frac{T}{K}\right)$ is more accurate. This gives the more exact value, 8.82×10^3 GHz. At 1/10th of this frequency, 8.82×10^2 GHz, from Fig. 1.7 (see statement of this problem), the prediction from the Rayeigh-Jeans and Planck relations agree. These give an value that is about 6% of the peak intensity.

For T=2.73 K, Eq. (1.25) (see statement of problem) gives a frequency of 160.3 GHz. From figure 1.1, the intensity is 10^{-18} W m^{-2} Hz^{-1} sterad^{-1}. Using this in Eq. (1.32), which is in MKS units, we have $(T_B = \frac{\lambda^2}{2k} I_\nu)$. Inserting numerical values, we have for the approximation to the Planck law:

$$T_B = ((0.00273)^2 \times 10^{-18})/(2 \times 1.38 \times 10^{-23}) \times 4\pi$$

The result is 3.39 K. Clearly, for mm wavelengths and low temperatures, the exact Planck law is needed.

19. One needs Eq. (1.37) of 'Tools', which is: $T = T_B(0) \times e^{-\tau} + T(1 - e^{-\tau})$. Then $T(\text{out}) = 1\,\text{K}\,e^{-0.1} + 300\,\text{K}\,(1 - e^{-0.1}) = 29.4\,\text{K}$. One should cool the cable, since this reduces the (useless, i. e. noise, rather than signal!) contribution of the cable itself to the output signal.

20. Use the equation of radiative transfer: $T_1(\text{out}) = T_s e^{-\tau} + T_1(1 - e^{-\tau})$. Then, $T_2(\text{out}) = T_s e^{-2\tau} + (1 - e^{-\tau})[T_1 e^{-\tau} + T_2]$. To minimize the output, the last term has to be as small as possible. This is achieved if the warmer cable, i.e., the one having T_1, is placed first.

21. Inserting values into Eq. (1.43), we have the following. The value of Boltzmann's constant in MKS units: $k = 1.38 \times 10^{-23}\,\text{J K}^{-1}$, and $1\,\text{J s}^{-1}$ to obtain P, power in Watts. For 1 mW, the value of T is 7.24×10^{19} K.

22*. From Eq. (1.40, in statement of this problem), we have

$$T_{B0} = (1 - r)T_0 + rT_s$$

The Black Body intensity per m^2 from the warm surface is given by $r \times \pi \times I_\nu$ where I_ν in MKS units is given in problem 13:

$$B_\nu = 1.47 \times 10^{-20} \times (\nu(\text{GHz}))^3 \times \frac{1}{e^{0.048\nu(\text{GHz})/T} - 1}$$

The effective temperature is calculated using:

$$J(T) = \frac{h\nu}{k} \times \frac{1}{e^{h\nu/kT} - 1}$$

we can calculate T_0 (Table 17.1).

Table 17.1 Effective temperatures and power for a reflector (problem 22)

ν (GHz)	$h\nu/k$ (K)	$J(T)$ (K)	$r \times T_0$ (W m^{-2} Hz^{-1})	$\pi\, r \times S_\nu$ (W m^{-2})
100	5.0	297.6	29.8	9×10^{-14}
1000	48.0	276.6	27.7	8×10^{-12}
10^4	480.0	121.5	12.2	4×10^{-9}

Although the effective temperature is slightly higher at 100 GHz, the losses are usually much lower. At 10^4 GHz, the effective temperature is lower, but the combined losses are usually much higher. More important is that the flux density and power in W m^{-2} (taking into account a 10% bandwidth) input is 10^5 higher at 10^4 GHz than at 100 GHz. This is important considering the fact that for bolometers, the fractional bandwidth is larger at higher frequencies. This bandwidth is typically at least 10% of the center frequency.

Chapter 18
Solutions for Chapter 2: Electromagnetic Wave Propagation Fundamentals

1. The solid angle for $\tan\theta = 20\,\text{km}/40\,\text{km}$, or $\theta=26.5°$. This is $\Omega = 2\pi \int_0^\theta \sin\theta\, d\theta = 2\pi(1 - \cos\theta) = 0.663$ steradians. By inserting the numbers, the flux is determined to be $2\ 10^{-10}\,\text{W}\,\text{m}^{-2}$. To reach the edge of a city with a radius of $20\,\text{km}$, this is $47.7\,\text{km}$ from the transmitter, with the required power flux the radiated power must be $P = 2 \times 10^{-10}\,\text{W}\,\text{m}^{-2} \times [(47.7 \times 10^3)]^2\,\text{m}^2 = 0.4\,\text{W}$. This is a very small quantity. The danger level is at a distance R, where the flux reaches $0.4\,\text{W}/R^2 = 10^{-2}\,\text{W}$, so $R = 6.3\,\text{m}$. This is a proposal put forth by *Facebook*.

2. (a) The source luminosity, L, is $L = 4\pi \times D^2 \times S_\nu$ For the values given, this is
$L = 4\pi \times 1.88 \times 10^3 \times 3 \times \left(10^{16}\right)^2\,m^2 \times (600\,\text{Hz}) \times \left(10^{-23}\,\text{W}\,\text{m}^{-2}\,\text{Hz}^{-1}\right)$
The result is $2.4 \times 10^{20}\,\text{W}$.
Using Eq. (1.34) (this is in the statement of problem 10 of chapter 1.),

$$10^3 \text{Jy} = 2.65\,T_B \times (10^{-3}/60)^2/18^2 = 2.16 \times 10^{-8}$$

the result is

$$T_B = 4.6 \times 10^{10}\,\text{K}$$

This is more than could be expected from any thermal process. At first, some speculated that this was a communication from an alien civilization. The more mundane solution is maser emission. This is a Λ-doublet line of the OH molecule.

3. From Eq. (2.64), this is: $v_{\text{phase}} \times v_g = c^2$ Thus the group velocity is $v_g = c\sqrt{1 - (\lambda_0/\lambda_c)^2}$ If $\lambda_0 = 1/2\lambda_c$, we have $v_g = c/\sqrt{2}$ and the phase velocity is $c \times \sqrt{2}$. The group velocity is the speed at which information can be transmitted, whereas the phase velocity is has no such interpretation.

© Springer International Publishing AG, part of Springer Nature 2018
T. L. Wilson, S. Hüttemeister, *Tools of Radio Astronomy – Problems and Solutions*, Astronomy and Astrophysics Library,
https://doi.org/10.1007/978-3-319-90820-5_18

4. From 'Tools', Appendix A, Eq. (A.24), the Fourier transform (FT) is

$$F(t) = \int_{-\infty}^{+\infty} e^{-\frac{\nu^2}{(\Delta\nu)^2}} e^{-i2\pi\nu t} d\nu \ .$$

From the relations for the Fourier Transform (Table A.2), it is found that

$$F(t) = \sqrt{\pi}\,\Delta\nu\,e^{-\pi^2(\Delta\nu)^2 t^2} \ .$$

Requiring the exponent to be equal to $-t^2/(\Delta t)^2$ yields $\Delta t = 1/(\pi\,\Delta\nu)$. Thus, the product of the widths in t and ν is $\Delta\nu\Delta t = 1/\pi$. This is a fundamental relation. In quantum mechanics, this is the 'uncertainty principle' governing joint uncertainties in momentum and position or phase and energy. See problem 4 in chapter 12.

5. This can be solved using the 'shift theorem' in Table A.2 of 'Tools':

$$f(x - a) = e^{-i2\pi as}$$

where 'a' is k_0. A more straightforward 'brute force' application of the integrals is also possible.
If $a(k) = a_0$ for $k_1 < k < k_2$, this is a square box in k space, and a factor $(\sin x/x)$ in time. See the solution to the convolution of two Gaussians in problem 6 of chapter 8.

6. Differentiating Eq. (2.67) in 'Tools' with respect to t (using the appropriate units, $[\text{cm}^{-3}\,\text{pc}]$ for the dispersion measure, $[\text{s}]$ for the pulse arrival time τ_D and $[\text{MHz}]$ for the frequency) gives

$$\dot{\tau}_D = 4.148 \times 10^3\,\text{DM}\,\frac{2\dot{\nu}}{\nu^3} \quad \rightarrow \quad \dot{\nu}[\text{MHz/s}] = 1.20 \times 10^{-4}\,(\text{DM})^{-1}\,\nu^3 \ .$$

7. (a) Set the dispersion bandwidth, $\Delta\nu_D$ equal to B, the bandwidth which leads to a pulse width of Δt. Then, using the relation from the last problem, we have $B/(\Delta t) = 1.2 \times 10^{-4}\,(\text{DM})^{-1}\nu^3$. Thus, $\Delta t[\text{s}] = 8.3 \times 10^3\,B\,\text{DM}\,(\nu^{-3}[\text{MHz}])$
(b) The dispersion measure of the ionosphere is DM $= (10^5\,\text{cm}^{-3}) \times (6.7 \times 10^{-13}\,\text{pc}) = 6.7 \times 10^{-8}\,\text{pc}\,\text{cm}^{-3}$. For $\nu = 100\,\text{MHz}$, $\Delta t[\text{s}] = 5.5 \times 10^{-10}\,B$. So even with a bandwidth of $10\,\text{MHz}$ the smearing is small compared to a typical pulse length of 10^{-6} s.

8.(a) Rewrite the result in Problem 7(a) with $(202)^3 \approx 8.3 \times 10^6$. So, $\Delta t[\text{ms}] =$ DM $(202/\nu[\text{MHz}])^3\,B$.
(b) The dispersion measure is DM $= (0.05\,\text{cm}^{-3}) \times (5.0 \times 10^3\,\text{pc}) = 250\,\text{cm}^{-3}\,\text{pc}$, so the pulse smearing at $400\,\text{MHz}$ is $\Delta t[\text{ms}] = 250 \times (202/400)^3\,B = 32.2 \times B$ ms. At $800\,\text{MHz}$, this is $4.0 \times B$ ms.

9. The dispersion measure is DM $= 10^3\,\text{cm}^{-3} \times 10\,\text{pc} = 10^4\,\text{cm}^{-3}$ pc. The pulses are separated by 3.3×10^{-2} s, so the smearing must be significantly less than this value. We require it to be at most $10\,\text{ms}$. From the result of Problem 8, we have

$\Delta t[\text{ms}] < 10 = 10^4\,\text{cm}^{-3}\,\text{pc}\,(202/1000)^3\,B$, or $B \le 0.12\,\text{MHz}$, where B is the bandwidth.

10. Use Eq. (2.71):

$$\Delta\tau(\mu s) = 4.148 \times 10^9 \left[\text{DM(cm}^{-3}\,\text{pc})\right]\left[\frac{1}{v_1^2} - \frac{1}{v_2^2}\right]$$

where v is in units of MHz. Then set $v_2 = \infty$, and set DM=30 and v_1=400 MHz, so obtain $\Delta\tau(\mu s)$=0.78 s. For v_1=1400 MHz, $\Delta\tau(\mu s)$=0.12 s

11. Use Eq. (2.73), where b is the bandwidth: $b_{\text{MHz}} = 1.205 \times 10^{-4}\,\frac{1}{\text{DM}}\,v^3(\text{MHz}) \times \tau_{\text{sec}}$ For v=1000 MHz and DM=30, have

$$b = (1.205 \times 10^{-4}/30) \times (1000)^3 \times (0.1 \times 10^{-6})$$

This gives 4.2×10^{-4} MHz, or 0.4 kHz.

12. Use Eq. (2.73), with v=1.4 $\times 10^3$ MHz and $\tau = 0.01$ s. Then

$$b = \left(1.205 \times 10^{-4}/295\right) \times (1400)^3 \times (0.01) = 11.2\,\text{MHz}$$

Chapter 19
Solutions for Chapter 3: Wave Polarization

1. Refer to Fig 3.3 of 'Tools', given here as figure 3.1. One must reverse the direction of the arrows on the circles in the upper and lower parts of this figure.

2. Eq. (3.39) from 'Tools' was given in the statement of this problem.
Use the definitions to determine that $S_1 = Q = E_1^2$, so the wave is linearly polarized, aligned along the North/South axis.

3. From the statement of this problem, the wave is linearly polarized at an angle of $45°$ with respect to the N/S axis.

4. Then $S_3 = V = S_0$. This is the opposite of the statement in 'Tools', p. 49, point 2. Also the value of V in point 1, p. 49 of 'Tools' needs a minus sign.

5. In the north, the wave is receeding Counter Clock Wise (CCW), it is Left Hand Circular, LHC.

6. (a) The rotation measure is defined as:

$$\text{RM} = 8.1 \times 10^5 \, (1\, G \times (10^5 \, \text{cm}^{-3}) \times (6.5 \times 10^{-13}) = 5.3 \times 10^{-2} \, \text{rad m}^{-2}$$

(b) For 100 MHz, we have $\Delta\psi = 0.48$ rad $= 27.3°$. For 1 GHz, $\Delta\psi = 0.27°$, and for 10 GHz, $\Delta\psi = 0.003°$.

(c) If the magnetic field is perpendicular to the line of sight, there is no effect, so the results in **(b)** are upper limits. Circular polarization is not changed by Faraday rotation.

(d) Inserting the numbers gives RM $= 8.1 \times 10^5$ (5×10^{-6} G) $\times$ (5 cm^{-3}) $\times$ (5×10^{-5} pc) $= 1.01 \times 10^{-3}$ rad m^{-2}. For 100 MHz, $\Delta\psi = 0.009$ rad $= 0.5°$. For 1 GHz, the effect is 1% of the result for 100 MHz ($\Delta\psi = 0.005°$), so no corrections are needed.

© Springer International Publishing AG, part of Springer Nature 2018
T. L. Wilson, S. Hüttemeister, *Tools of Radio Astronomy – Problems and Solutions*, Astronomy and Astrophysics Library,
https://doi.org/10.1007/978-3-319-90820-5_19

7. Using the same relation as above, the rotation measure is RM $= 2.2 \times 10^2 \, \mathrm{rad \, m^{-2}}$. For 100 MHz, $\Delta\psi = (3 \, \mathrm{m})^2 \, \mathrm{RM} = 2 \times 10^3 \, \mathrm{rad} = (1.1 \times 10^5)°$. This is a very large rotation measure. For 1 GHz, the result is 1% of this value, which is smaller, but still large, with $\Delta\psi = 19.7 \, \mathrm{rad}$.

8. (a) Since the electric field must be zero in the metal, at the instant when the incoming wave reaches the metal surface, there must be a wave with the opposite sense of circular polarization leaving the surface, so that the sum of the two waves cancel. Thus, the sense of circular polarization of the reflected wave is opposite to that of the incoming wave.

(b) For a linearly polarized wave, the polarization is unchanged, but the phase of the outgoing wave is 180° different.

9. Using the definitions of RM and DM and taking the ratio, we get $B_\parallel = 1.23 \times 10^{-6} \, \mathrm{RM/DM} = 3.0 \times 10^{-6} \, \mathrm{G} = 3 \, \mu\mathrm{G}$. If the field perpendicular to the line of sight is also $3 \, \mu\mathrm{G}$, the total vector sum is $4.2 \, \mu\mathrm{G}$.

10. We start with Eq. (3.41), which is: $V(t) = \int_0^\infty a(\nu) e^{i[\phi(\nu) - 2\pi\nu t]} \, d\nu$. We then set $a(\nu) = a_0$, $\phi(\nu) = \phi_0$ and make the upper limit of the integral equal to $0.1\nu_0$. Then $V(t) = \int_0^{0.1\nu_0} a_0 \, e^{i[\phi_0 - 2\pi\nu t]} \, d\nu$. This gives:

$$V(t) = -a_0 \, e^{i\phi_0} \, e^{-\pi i(0.1\nu t)} \frac{1}{2\pi \, it} \times [e^{-\pi i(0.1\nu)t} - e^{\pi i(0.1\nu)t}]$$

This is:

$$V(t) = -a_0 \, e^{i\phi_0} \, e^{-\pi i(0.1\nu t)} \times \frac{\sin[0.1\nu_0 \, \pi \, t]}{2\pi \, it}$$

11. Here we show how to carry out the integration with all of the mathematical details. The Fourier Transform is F. This is a function of time, whereas the input is a function of frequency, ν. Then, from Appendix A, we have:

$$F(s) = \int_{-\infty}^\infty f(x) e^{-i2\pi \, xs} ds$$

We drop the range of integration, and substitute t for s and ν for x. Then:

$$F(t) = \int e^{-i2\pi \nu t} \left(a_0 e^{-(\nu - \nu_0)^2/\Delta\nu^2} \right) d\nu$$

change $\nu - \nu_0$ to ν'. Then $d\nu' = d\nu$, since ν_0 is a constant. The goal is transform the exponent into a quantity which is a square. This is done by adding the term $(2\pi)^2 t^2$. Then the exponent is renamed ν'' and we have:

$$F(t) = e^{(2\pi)^2 t^2} \int e^{-(k''/\Delta\nu)^2} dk''$$

The integral is a Gaussian, with the result $\sqrt{\pi \Delta\nu^2}$. The standard interpretation is as given in problem 4 of chapter 2, namely a large width in frequency requires a small width in time.

12. The gyrating electrons emit synchrotron radiation, which has a high degree of linear polarization. The circular orbit of an electron at the magnetic equator is perpendicular to the movement of an electron at the magnetic poles. Thus, a telescope with a small beam moving from the equator to the north or south pole will measure a smooth change in linear polarization angle by 90°.

Chapter 20
Solutions for Chapter 4: Signal Processing and Receivers: Theory

1. (a) We must evaluate the integral $\int_{-\infty}^{+\infty} A e^{-x^2/2\sigma^2} dx = 1$. The standard approach is to evaluate the square of this integral in terms of the variables x and y. Then we have $A^2 \int_{-\infty}^{+\infty} e^{-x^2/2\sigma^2} dx \int_{-\infty}^{+\infty} e^{-y^2/2\sigma^2} dy = 1$. Now transform from rectangular to two-dimensional polar coordinates, so that

$dx dy = d\theta dr$, and $x^2 + y^2 = r^2$.

Then $A^2 \int_0^{2\pi} \int_0^{+\infty} e^{r^2/2\sigma^2} = 1$. The result is $A = 1/\sqrt{2\pi\sigma^2}$.

(b) $m = \langle x \rangle = 0$, since $\int_{-\infty}^{+\infty} x e^{-x^2/2\sigma^2} dx = 0$.

For σ, we need $\langle x^2 \rangle = (1/\sqrt{2\pi\sigma^2}) \int_{-\infty}^{+\infty} x^2 e^{x^2/2\sigma^2} dx$. We must use 'integration by parts', that is $\int u dv = uv - \int v du$, with $v = -\sigma e^{x^2/2\sigma^2}$ and $u = x$, so that $dv = x e^{-x^2/2\sigma^2}$.

Then $\langle x^2 \rangle = \sigma^2 \int_{-\infty}^{+\infty} e^{x^2/2\sigma^2} dx = (1/\sqrt{2\pi\sigma^2})\sigma^2\sqrt{2\pi\sigma^2} = \sigma^2$.

$\langle x^3 \rangle = 0$, while $\langle x^4 \rangle = (1/2\pi\sigma^2) \int_{-\infty}^{+\infty} x^4 e^{-x^2/2\sigma^2} dx$. This latter expression can be evaluated by the use of 'integration by parts', with $u = x^3$ and $dv = x e^{-x^2/2\sigma^2}$. Then $v = -\sigma^2 e^{-x^2/2\sigma^2}$ and $du = 3x^2$. The surviving term is $(3\sigma^2/2\pi\sigma^2) \int_{-\infty}^{+\infty} x^2 e^{-x^2/2\sigma^2} dx = (3\sigma^2) \times (1/2\pi\sigma^2) \times (\sqrt{\pi\sigma^2}/2) = 3\sigma^4$.

(c) From the results of part **(b)**, we find that $\sqrt{x^4} = \sqrt{3} \langle x^2 \rangle$. Then we can sample the voltage, square this and square again. Then one can investigate the σ values of these results and compare to the theoretical result given here.

2. Use Eq. (4.11) in 'Tools', namely $R_T(\tau) = E_T\{x(s) x(s+\tau)\} = E_T\{x(t-\tau) x(t)\}$. In this case, since the function is periodic, use the period of the sine wave. Then $R(\tau) = \int_0^{1/\nu} (A \sin 2\pi\nu t)(A \sin 2\pi\nu \times (t-\tau)) dt$. Expanding the time-delayed term, we have

$R(\tau) = A^2 \int_0^{1/\nu} \left[(\sin 2\pi\nu t)^2 \cos(2\pi\nu\tau) + \sin(2\pi\nu t) \cos(2\pi\nu t) \sin(2\pi\nu\tau) \right] dt$.

Substituting $x = 2\pi\nu$, we find that the second term is zero, while the first term is $R(\tau) = A^2 (1/2\pi\nu)(\pi)(\cos 2\pi\nu\tau) = (A^2/2\nu) \cos 2\pi\nu\tau$.

© Springer International Publishing AG, part of Springer Nature 2018
T. L. Wilson, S. Hüttemeister, *Tools of Radio Astronomy – Problems and Solutions*, Astronomy and Astrophysics Library,
https://doi.org/10.1007/978-3-319-90820-5_20

For the cosine, the result is the same. In the frequency domain, the power spectrum is a single spike at $\nu = 1/2\pi\tau$. This process is illustrated in fig. 20.1

3. Use Eq. (4.7) in 'Tools', namely: $X(\nu) = \lim_{T\to\infty} \int_{-\frac{T}{2}}^{\frac{T}{2}} x(t)e^{-2\pi i\nu t}dt$; the Fourier transform extends to $\pm\infty$, but the function is zero at $\pm\tau/2$. Then

$v(\nu) = \int_{-1/2\tau}^{+1/2\tau} f(t)e^{-2\pi i\nu t}dt = -(A/2\pi i\nu)\left(e^{-2\pi i\nu\tau/2} + e^{2\pi i\nu\tau/2}\right)$
$\qquad = A\tau\left(\sin(\pi\nu\tau)/\pi\nu\tau\right).$

The spectral power density is
$S(\nu) = v(\nu)^2 = A^2\tau^2\left(\sin(\pi\nu\tau)/\pi\nu\tau\right)^2.$

The autocorrelation function, $R(\tau)$, is calculated as in Problem 2, this chapter. A graphical method to obtain this is to slide the two rectangles past each other, stepwise, summing the overlap at each step. The result is a triangle. The Fourier transform of a triangular function is $(\sin x/x)^2$ as obtained above. Fig. 4.1 in 'Tools', is here figure 20.2. This shows a sketch of this process in terms of voltages and power (i.e. power). A more general relation describes the relation of autocorrelation and Fourier Transforms.

4. This is $v(\nu) = \int_0^\tau f(t)e^{-2\pi i\nu t}dt$ which equals $-A/2\pi i\nu\left(e^{-2\pi i\nu\tau} + 0\right)$. This is: $v(\nu) = A/(2\pi i\nu)\left(e^{-\pi i\nu\tau}\right)(\sin\pi\nu\tau/\pi\nu\tau)$. This is the function found in Problem 5, this Chapter, shifted by the factor $e^{-\pi i\nu\tau}$.

5. See fig. 20.2 for details. To graphically produce an autocorrelation, we must first invert the shape about the f axis, then calculate the overlap. Then we sum over the

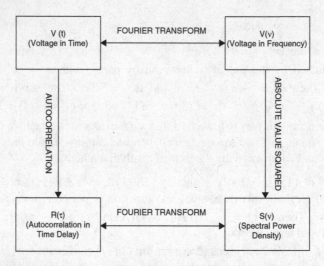

Fig. 20.1 A sketch of the relation between the voltage input as a function of time, $V(t)$, and frequency, $V(\nu)$, with the autocorrelation function, ACF, $R(\tau)$, where τ is the delay, and corresponding power spectral density, PSD, $S(\nu)$. The two-headed arrows represent reversible processes. This a specific example of the more general relation between autocorrelation (on the left) and Fourier Transforms (problem 2)

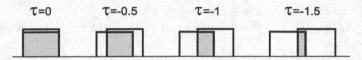

Fig. 20.2 Graphical display of the correlation of two square-shaped functions as a function of offset. This process is treated in Problem 5

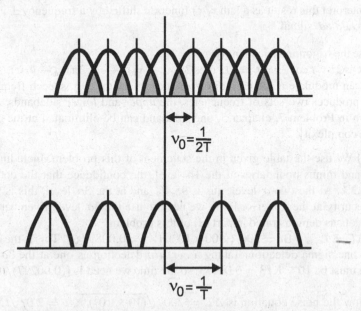

Fig. 20.3 This refers to Problem 6. *Above* is the distribution in frequency space in which the time sampling interval is too long for the frequency spread of the function. There is an overlap of frequencies of the sampled function, shown as shaded areas. This is an example of *aliasing*. *Below* the time sampling interval is halved, so aliasing is avoided

overlapping area, and plot this as a function of t_0. We must repeat the process for a series of t_0 values, both positive and negative. For those with enthusiasm, apply this process to the figure on the left side of Fig. 4.10 in 'Tools'.

6. In frequency space, the convolution is a multiplication of the Fourier Transformed functions. The result is a series of identical $F(\nu)$ structures. If the maximum frequency content of $F(\nu)$ is large enough there will be an overlap of adjacent functions, which is referred to as *aliasing*. Clearly, halving the sampling rate causes an overlap of the sampled data (figure 20.3).

7. Use the analysis in problem 4, this chapter, to obtain
$v(\nu) = (A/2\pi i\nu)\left(e^{-\pi i\nu\tau}\left(\sin(\pi\nu\tau)/\pi\nu\tau\right) + e^{\pi i\nu\tau}\left(\sin(\pi\nu\tau)/\pi\nu\tau\right)\right)$. Combining
terms, we have
$$v(\nu) = (A/2\pi i\nu)\cos(\pi\nu\tau/2)\left(\sin(\pi\nu\tau)/(\pi\nu\tau)\right)$$
$$= (A/4\pi^2 i\nu^2\,\tau)\left(\sin(3\pi\nu\tau/2) - \sin(\pi\nu\tau/2)\right).$$

We interpret this result as a $(\sin x/x)$ function shifted by a frequency $\pm\nu/2$. This is
a *modulated* signal.

8. Use the trigonometric identity:
$$y = \cos 2\pi\nu_c t\cos 2\pi\nu_s t = 1/2\left(\cos 2\pi(\nu_c + \nu_s)t + \cos 2\pi(\nu_c - \nu_s)t\right).$$
One can modulate a carrier radio frequency, ν_c, with, e. g. speech frequencies ν_s.
This produces two sets of frequencies, the *upper* and *lower* sidebands. As will be
shown in Problem 7, chapter 5, one sideband can be eliminated at the expense of
more complexity.

9. (a) We use the table given in the statement of this problem. Including both the
plus and minus boundaries at the 1σ level, the confidence that the correct result
is 68.3%. At the two σ level, this is 95.4% and at the 3σ level, this is 99.7%. So
for security at the 1% error level, we need to use the 3σ level. Even better is a 4σ
criterion, as demonstrated in part (d) of this problem.
(b) $\Delta T = T_{sys}/\sqrt{Bt} = 200/[(500\times 10^6)t]^{1/2} = 0.0089/\sqrt{t}$. This is the expression
for a one-sigma detection. Taking a *certain* detection as one at the 3σ level, the
noise must be 10^{-2} K$/3 = 0.00333$, so the time we need is $[(0.00297)/(0.01)]^2$, or
7.2 s.
(c) Now the noise equation is $\Delta T = 200/\sqrt{(10\times 10^3)\times t} = 2.0/\sqrt{t}$. Using the
same 3σ value as in part **(c)**, we find that 100 h are needed. Clearly, the bandwidth is
an important factor. If one can assume that the line is only in *emission*, one might
believe that the integration time is 1/4 of this value, since only one half of the value
outside the Gaussian integrated area is needed. However, this statement makes use
of additional information, which is not normally used to argue for the integration
time needed for a detection.
(d) Using Table 20.1, at the 1σ we have 317 channels with spurious detections
and at the 2σ level, 46 channels. At the 3σ level, there will be only three spurious
channels. Even better is the use of a criterion that sets the reality of a detection at
the 4σ level. Then, there are only two spurious features.
(e) The use of additional information could allow one to reduce the integration time.
Restricting a feature to positive-only values would eliminate one half of the area

Table 20.1 Y factor and receiver noise temperature

T_{RX} (K)	Y from Eq. (4.11)	T_{RX} from Eq. (4.34)	% difference
1000	1.24	810	19
100	2.0	136	36
20	3.3	17	15

outside the Gaussian curve's boundary. Then the integration time would be a factor of 4 lower. However, see the statement at the end of the solutions for part (c).

10*. (a) Differentiate the expression for σ with respect to f. Then set this to zero to minimize the uncertainty in the combined sigma: the result is $f = \sigma_2^2/(\sigma_1^2 + \sigma_2^2)$. Substitute into the expression for $\bar{x}$. The results are given in the question.

(b) If we take $\sigma_1^2 \sim 1/t$, we obtain $\bar{x} = (1/(t_1 + t_2))(t_1 x_1 + t_2 x_2)$. This is as it must be to have an average which we intuitively expect.

11. If the receiver contributes no noise, $P_{out} = G P_{in}$ or $F = 1$. If $F = 1$, the receiver noise is zero. If we put a room temperature load ($T = 290\,\text{K}$) on the input, and if output power is doubled, must have $T_{rx} = 290\,\text{K}$. If $F = 1.2$, $T_{rx} = 58\,\text{K}$, for $F = 1.5$, $T_{rx} = 145\,\text{K}$.

12*. Use the analysis in Section 4.2.1; the process is analogous. This analysis for the square law detector is as follows. The steps refer to figure 4.4 in the statement of this problem.

$P_2 = v_2^2 = \sigma^2 = kT_{sys}G\Delta v$

$< v_3 > = < |v_2| > = \int_{-\infty}^{\infty} |v_2| p(v_2) dv_2 = 2 \int_0^{\infty} v_2\, p(v_2) dv_2$.

Set $v_2/2\sigma = x^2$

$< v_3 > = 4\sigma^2 (1/\sqrt{2\pi\sigma^2}) \int_0^{\infty} x e^{-x^2} dx = (\sqrt{2/\pi})\sigma$

$< v_3^2 > = < v_2^2 > = \sigma^2$, so $\sigma_3^2 = < v_3^2 > - < v_3 >^2 = \sigma^2 - (\sqrt{2/\pi}\,\sigma)^2 = (1 - (2/\pi))\sigma^2$

But $< v_3 > = < v_4 > = (\sqrt{2/\pi})\sigma = ([2/\pi] kGv[T_s + T_R])^{1/2}$

If $T_s \ll T_R$, have for the signal

$< v_4 >_{signal} = (2/\pi kGvT_R[1 + T_s/T_R])^{1/2} = (kGv/2\pi T_R)^{1/2} T_s$

$\sigma_4^2 = \sigma_3^2/2v\tau = \sigma^2(1 - 2/\pi)/2v\tau$.

To relate the σ_4 to temperature, need to form

$\Delta < v_4 >_{signal} / \Delta T_{signal} = (2 k B G \Delta v/2\pi T_R)^{1/2}$.

Then, for $T_R \gg T_s$, have

$< \sigma_4 > = \left(\dfrac{(1 - 2/\pi)(kG\Delta v T_R)}{\sqrt{2}\Delta v\tau} \right)$.

Since $< v_4 > = < v_3 >$, and using

$\Delta T_{RMS} = \dfrac{\sigma_4}{(\Delta < v_4 >/\Delta T_s)}$, we have

$$\Delta T_{RMS} = \frac{\sqrt{\pi - 2}\, T_R}{\sqrt{v\tau}} = \frac{1.07 T_R}{\sqrt{v\tau}}$$

This has slightly less sensitivity than the square law detector system, but more seriously, we can have a linear response only if the signal intensity is much less than the receiver noise.

13. Check three values in Fig. 4.11 from Tools, figure 4.5 in the statement of this problem.
The conclusion should be that the plot is illustrative, but that the Eq. (4.11) is more accurate.

14.(a) The receiver noise is not changed since the signal and the noise are in both ν_u and ν_l

(b) The receiver noise is twice as large, since the signal is in one sideband but the receiver noise is in both.

(c) If the response is twice as large, but the receiver noise is the same in both sidebands, the signal-noise ratio is $2\,S_u/(T_{rx_u} + T_{rx_l}) = S_u/T_{rx_u}$

15. Eq. (4.56) states:

$$\frac{\Delta T}{T_{sys}} = K\sqrt{\frac{1}{\Delta\nu\,\tau} + \left(\frac{\Delta G}{G}\right)^2}.$$

Differentiating this with respect to τ and setting the result equal to zero, we have the term: $-1/\Delta\tau^2 + \gamma_1 = 0$. The solution is $\tau^2 = 1/(\gamma_1 + \Delta\nu)$. This gives $\tau = 1/\sqrt{\gamma_1\,\Delta\nu}$

16. For the on-source measurement consisting of ten samples, the signal at the position of the source is 10, while the noise at the position of the source is $\sqrt{10}$. Thus the signal-to-noise (S/N) ratio improves as $\sqrt{10}$ At the position of the reference, the signal is 1000, while the noise is $\sqrt{1000}$. Thus the S/N ratio is $\sqrt{1000}$. When the difference of the on-source and reference is taken, the noise in this result is the square root of the sum of the squares of the noise in the on-source and reference, so the S/N ratio has improved by $(1 + 1000)/(\sqrt{1 + 1000}) = \sqrt{1 + 1000}$. For an on-source combined with an off-source measurement of equal time, the result would be $\sqrt{2}$. This has no advantage for a single off-on pair. However, for a map of an extended region, there are advantages. If one takes 1000 off-source samples and then 1000 on-source samples at different positions, the on-off combination gives an advantage of $\sqrt{1001/2}$. In the more general case, if the number of on-source samples is 1 N and the number of reference is 1000 N, the noise in the difference of on-minus-off is $N\sqrt{1 + 1000}$. In contrast, the S/N ratio is 1001. So, for mapping an extended source, the combination of 1000 samples on the reference position with 1000 different positions on the source gives a great advantage in S/N ratio when compared to a single on-off pair.

17. Use the relation $T_{min} = h\nu/k$. Then we have:
for $\nu = 115.271$ GHz, $T_{min} = 5.5$ K,
for $\nu = 1000$ GHz, $T_{min} = 48$ K, and
for $\nu = 10^5$ GHz, $T_{min} = 4800$ K.

18. The $\Delta\nu$ solutions can be divided into two parts. First is the predetection bandwidth. After that, the smoothing function.
For the numerator, we square the result after carrying out the integral. For the rectangular pass band, the step function gives $\int_{\nu_0-1/2\Delta}^{\nu_0+1/2\Delta} d\nu$, which equals Δ: This is squared to give $2\,\Delta^2$. For the denominator, the square of the step function is the function, so is δ. Taking the factor of 1/2 into account, the ratio is thus Δ (see also problem 6). For the single tuned circuit, the numerator is $\int_{-\infty}^{\infty} \frac{1}{[1+(|\nu|-\nu_0)^2/\Delta)^2]}d\nu$.

The correlation integral results in an arctangent, which is π. (This integral, and all of those handled here, are to be found in 'Table of Integrals, Series and Products', 7th edition by I.S. Gradshteyn and I.M. Ryzhik (Jeffrey and Zwillinger, eds., Elsevier Academic Press 2007, in sections 3.249 for numerator 2.141 for denominator)). The numerator is multiplied by 1/2, so the autocorrelation is halved to give the result. The Gaussian integral for the numerator gives $2\,\Delta\,\sqrt{2\pi}$, which is squared. The denominator gives $\Delta\sqrt{\pi}$. So the ratio is $2\Delta\,\sqrt{\pi}$. The factor of 1/2 provides the final result.

For the smoothing functions, the numerators are all unity. The integral in the denominator for the 'running mean' is $\int(\sin x/x)^2 dx$ which is related to the Dirichlet integral. This integral has the value π. Changing the variable from $2\pi T\nu$ to x to simplify the integration yields the result in the table. For the rectangular passband, the integral is from 0 to ν_0 and $-\nu_0$. The result is $2\nu_0$. Since this is in the denominator, the result is $1/2\nu_0^{-1}$. For the Gaussian integral, the result is $\sqrt{2\pi}$. Since the smoothing is the inverse, this is $\frac{1}{\sqrt{2\pi}}$.

Inserting these into the relation $\frac{1}{\sqrt{\Delta\nu\,\tau}}$, one can assess the effect of using different bandwidths and smoothing functions.

Chapter 21
Solutions for Chapter 5: Practical Receiver Systems

1. Fig. 5.13 is shown in the statement of this problem as figure 5.1. This is a log-log plot. The value of the y-axis at 10 GHz is 4.3 K, while at 1000 GHz, this is 430 K. Thus the slope is 0.43 K/GHz. The minimum noise in a coherent receiver is $h\nu/k$ or 0.048 K/GHz, so to within the error of estimating the value of the slope in the plot, it is consistent with the report that this line is $10 \times h\nu(\text{GHz})/k$.

2. Use MKS units, and set the coupling efficiency, ε to unity: NEP=10^{-16}W Hz$^{-1/2}$= $2(1.38 \ 10^{-23} \ T_n \ \sqrt{50 \times 10^9})$ so we find T_n=16 K. Calculate the RMS noise from ΔT_{rms}= $T_n/\sqrt{B}$=7.2 10^{-5} K.
For the coherent receiver, $\Delta T_{RMS} = 50K/(2 \times 10^9 \tau)^{1/2} = 1.1 \times 10^{-3}/(\tau)^{1/2}$. To reach the RMS noise obtained with the Bolometer in 1 s, but one must integrate with the coherent receiver for 238 s.

3. If the sideband ratio is unity, the single-sideband noise is twice the receiver noise in the plot, since the receiver and sky noise enter in both the upper and lower sideband, whereas the signal enters from only one sideband.

4. For a 2 stage amplifier system, $T_{sys} = T_{stage1} + T_{stage2}/G_{stage1}$, with G_{stage1} being the gain of stage 1. For this problem, the gain, G_{stage1}, is 1000, so the system noise, $T_{sys} = 4 \text{ K} + (1000 \text{ K}/1000) = 5 \text{ K}$. Thus, the second stage contributes only 20% of the total system noise even though the noise of this component is 50 times higher than the noise of the first stage.

5. (a) The separation of the upper and lower sidebands is twice the IF frequency. Thus this $\nu(\text{IF}) = [(115 - 107)/2] \text{ GHz} = 4 \text{ GHz}$. The L. O. frequency is exactly between the sidebands, so $\nu(\text{LO}) = 111 \text{ GHz}$.

(b) If ν(signal) is higher than ν(LO), the IF frequency moves down, since the oscillator frequency is $\nu(\text{IF}) = \nu(\text{signal}) - \nu(\text{LO})$. We know that ν(signal) is fixed, and if ν(LO) is increased, ν(IF) will decrease, so the line must be in the upper sideband.

© Springer International Publishing AG, part of Springer Nature 2018
T. L. Wilson, S. Hüttemeister, *Tools of Radio Astronomy – Problems and Solutions*, Astronomy and Astrophysics Library,
https://doi.org/10.1007/978-3-319-90820-5_21

6. This is a more complex variant of the previous problem. For the lower sideband, $\nu(\text{IF}) = -\nu(\text{signal}) + \nu(\text{LO})$, so when $\nu(\text{LO})$ is increased, the line will move to higher frequency.

After the first mixer, the output from the signal band will move to lower frequency, if the L.O. to the first mixer is increased. After mixing, this line is at a lower frequency, so the frequency offset between the line (in the IF) and the second mixer is larger. So after the second mixing, the line moves to a higher frequency in the second IF.

7. Along the lower path in fig. 5.3, shown in the statement of this problem, the mixer output is shifted in phase by $\pi/2$, so this is $\cos(\omega t - \pi/2) = \sin \omega_s t$. The mixer in the lower path produces an output
$[\sin \omega_s t]^2 + [\sin \omega_c t]^2 + \sin \omega_c t \, \sin \omega_s t$. The upper path produces
$[\cos(\omega_s t)]^2 + [\cos(\omega_c t)]^2 + \cos(\omega_c t) \cos(\omega_s t)$. Adding these results, we have
$2 + \sin(\omega_c t) \sin(\omega_s t) + \cos(\omega_c t) \cos(\omega_s t) = \cos(\omega_c - \omega_s)t$.
This is the lower sideband signal. To obtain the upper sideband signal, subtract instead of summing.

8. The power amplification is 10^{18}, or 180 db. The output power will vary by 0.1%. One searches for a source which is 0.1 K. The gain fluctuations affect the total power output, receiver noise plus source noise. Thus, these fluctuations, 1 K, exceed the source intensity. In order to detect a source, one must switch between signal and a reference faster than once per minute to stabilize the output.

9. From Eq. (1.37) in 'Tools', we have
$T_s(\text{out}) = T(\text{signal})e^{-\tau} + T(\text{cable})(1 - e^{-\tau})$
This is the *signal* phase. For the reference phase, have
$T_r(\text{out}) = T(\text{cable})(1 - e^{-\tau})$.
The difference is the signal, S,
$S = T_s(\text{out}) - T_r(\text{out}) = T(\text{signal})e^{-\tau}$,
while the noise is $N = T_n = \sqrt{2}\,T(\text{cable}) \times (1 - e^{-\tau})$,
where the factor of $\sqrt{2}$ is the result of subtracting two noisy quantities. The signal-to-noise ratio will increase as $\sqrt{t}$ if only random effects are present.

10. $\Delta T_{\text{RMS}}/T_{\text{sys}} = 1/\sqrt{B\tau} = 0.001$; solving this, $\tau = 0.045$ s. For the instability given, we have $\Delta T_{\text{RMS}}/T_{\text{sys}} = 1/\sqrt{2 \times 10^{-9}/\tau + 10^{-6}\tau}$. The terms are equal after 0.045 s. One should switch much faster than 22 Hz to reduce the effects of gain instabilities.

11. For the cheaper receiver, the system noise temperature, including the including sky noise contribution, will exceed 150 K, while for the more expensive receiver the system noise temperature will exceed 110 K. For these systems, the ratio of the integration times to reach the same signal-to-noise ratio is a factor 1.85 (i. e. the square of the total receiver noise). If one can accept an extra factor of 1.85 in time, *use the cheaper receiver.*

12. (a) Three equally spaced samples taken within one period of a wave are needed to characterize a sine or cosine wave. In principle, two samples should characterize such a wave, if one is at a maximum and the other at a minimum, but these could be at zero-crossing points, and would not lead to a unique characterization of the spectrum.

(b) From the product of time and bandwidth, we have 108. Then $\sqrt{10^8} = 10^4$, we have $\Delta T_{RMS}/T_{signal} = 10^{-4}$, that is the signal-to-noise ratio is 10 000 or more everywhere. We need samples every 13′, or 21 samples per square degree. For the entire sky, this is $(41,252) \times (20) = 8.4 \times 10^5$ samples. At 10^s per sample, this is 2.34×10^3 h, or 98 days.

(c) For the 5 GHz survey, $\Delta T_{RMS} = 7.1 \times 10^{-4}$, using a total power receiver. Comparison switching is needed to obtain stability, so the noise is twice as large, 1.4×10^{-3} K. For three samples per beam, the spacing must be 0.8′, or 5625 samples per square degree. Over the whole sky this is 2.34×10^8 samples, which requires 6.5×10^5 h of time, or 74.2 years. Clearly, the 5 GHz survey must be carried out with a multi-beam receiver.

Chapter 22
Solutions for Chapter 6: Fundamentals of Antenna Theory

1. Eq. (6.49) is

$$\hat{S} = \Sigma_{n=0}^{N} e^{i\,k\,n\,D\,\sin(\phi)}$$

Set $e^{i\,k\,n\,D\,\sin(\phi)} = q$. Then the difference $\hat{S} - q \times \hat{S}$ is $1 - q^{N+1}$, since only the first an last terms of the series do not cancel. Solving for $\hat{S}$, we have: $\hat{S} = \frac{1-q^{N+1}}{1-q}$. Next convert q to the quantity $e^{-i\,k\,D\,\sin(\phi)}$ Then use the relation for $\sin x = 1/2 \times (e^{ix} - e^{-ix})$, to convert to trigonometric terms. Then factor a term $e^{-i\,(N/2)\,k\,D/2\,\sin(\phi)}$ out of the sum to convert to: The sum is

$$\hat{S} = e^{i\,k\,D\,\sin(\phi)} \times e^{-i\,N\,k\,D/2\,\sin(\phi)} \times \left[\frac{\sin\frac{k\,N\,D}{2}\sin(\phi)}{\sin\frac{k\,D}{2}\sin(\phi)} \right]$$

Square this term to obtain the power pattern.

$$|\hat{S}|^2 = \left[\frac{\sin(\frac{k\,N\,D}{2}\sin(\phi))}{\sin(\frac{k\,D}{2}\sin(\phi))} \right]^2$$

Since the imaginary factors cancel in the square. To obtain Eq. (6.64), allow N to increase, D to decrease, so that $N \times D = L_x$ the length of the aperture, is constant. In addition, the $\sin\phi$ term becomes a unit vector, l. Then $|\hat{S}|^2$ becomes the normalized power pattern in the x direction, as given by a simplified form of Eq. (6.64):

$$P_n(l) = \left[\frac{\sin(\pi l\,L_x/\lambda)}{\pi l\,L_x/\lambda} \right]^2$$

© Springer International Publishing AG, part of Springer Nature 2018
T. L. Wilson, S. Hüttemeister, *Tools of Radio Astronomy – Problems and Solutions*, Astronomy and Astrophysics Library,
https://doi.org/10.1007/978-3-319-90820-5_22

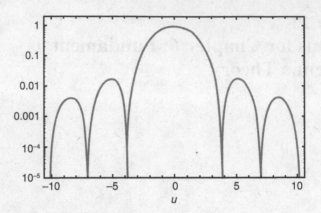

Fig. 22.1 This is Figure 6.10 from 'Tools'. It is used for problem 2 of this chapter. It is the power pattern for the unblocked, uniformly illuminated aperture, showing the main beam, and near and far side lobes. The vertical axis is normalized power, the horizontal axis is the parameter u. The first null occurs when u has the value 3.83. The full width between nulls is 2.44 rad, which is equal to λ/D

2. Fig. 6.10 of 'Tools', shown here as figure 22.1, is the power pattern of an unblocked, uniformly illuminated circular aperture. The sidelobes can be reduced by underilluminating the aperture, but there will still be sidelobes at some level. Thus, the report is false.

3. (a) The number of wavelengths is $D/\lambda = (0.5\,\text{cm})/(5 \times 10^{-5}\,\text{cm})$=10,000. For a 100 m radio telescope, the number of waves at $\lambda = 2$ m is 50, whereas at $\lambda = 2$ cm, this is 5000. For the ALMA 12 m antenna at $\lambda = 1$ cm, the number is $\lambda = 1200$, at $\lambda = 3$ mm, the number is 4000, and at $\lambda = 0.3$ mm, this is 40,000. The treatment of all of these systems must follow diffraction, but if the value of D/λ is more than 1000, one can use ray tracing and then correct the results for diffraction. For D/λ less than 1000, the treatment must follow diffraction strictly.

4. Eq. (6.53) gives the expression for the E field. This is:

$$E_\vartheta = \left(-\,\mathrm{i}\,\frac{I\,\Delta l}{2\lambda}\,\frac{1}{r}\right)\,\mathrm{e}^{\mathrm{i}\,(k\,r - \omega t)}\,\hat{S}$$

where $\hat{S}$ is given in problem 1 of this chapter. Setting $l = \sin\phi = 0$, the first order expansion of the terms in the square brackets in $\hat{S}$ becomes: $\hat{S} = N$, where N is the number of dipoles. The power pattern is the square of the E field, so the power from N dipoles is N^2 that of a single dipole.

5. (a) $\theta = 1.02 \times \lambda/D = (5 \times 10^{-5}\,\text{cm})/(0.3\,\text{cm}) = 1.7 \times 10^{-4}\,\text{rad} = 0.6'$.
(b) Put all units in centimeters:
$\theta = 1.02 \times (2\,\text{cm}/(100) \times 100\,\text{cm}) = 2 \times 10^{-4}\,\text{rad} = 0.7'$. For the Very Large Array, with $= 27$ km, this is
$\theta = 1.02 \times (2\,\text{cm}/(27) \times 1000 \times 100\,\text{cm}) = 7.4 \times 10^{-7}\,\text{rad} = 0.15\,\text{arcsec}$.

(c) This is a problem concerned with units. We start with $\theta = 1.02 \times \lambda/D$, the units of λ and D must match; choose meters. Then for λ in mm, need a factor of 10^{-3}, for D in km, need a factor of 10^3. So:

$$\theta = 1.02 \times \left[\left(10^{-3} \times \lambda\right)\right] / \left[\left(10^3 \times D\right)\right] \times 206, 265$$

so

$$\theta = 0.206 \frac{\lambda(\text{mm})}{D(\text{km})}$$

6. (a) $\theta = (26\,\text{cm})/(2\,\text{m}) = 0.13\,\text{rad} = 7.4°$.

(b) $R = c/3\,(\Delta l/\lambda)^2 = 10^8\,(0.003/0.20)^2 = (10^8\text{m}) \times (0.5\,\text{cm}/20\,\text{cm})^2 = 3.75 \times 10^4$ ohms.

(c) From Eq. (6.43) in 'Tools', in MKS units,
$P_{\text{total}} = 2c/3[I\Delta l/2\lambda]^2 = (2 \times 10^8\,\text{m s}^{-1})[(0.5\,\text{A}) \times 0.003\,\text{m})/2(0.2\,\text{m})]^2$
$= 7.8 \times 10^3$ W.

7. Total Energy $= (10^{-16}\,\text{W}) \times (40\,\text{telescopes}) \times (57\,\text{years} \times 3.14 \times 10^7\,\text{s/year}) = 7.2\,10^{-6}$ Joules $= 7.2$ ergs. The energy gained by $1\,\text{g}$ in falling $2\,\text{cm}$ is E=mgh or E=(1 g)(980 cm s^{-2})(2 cm)=1960 ergs. So the radio telescopes have received much less energy than gained by an ash falling 2 cm in the earth's gravity.

8. (a) Figure 6.11 from 'Tools' is given in the statement of this problem as figure 6.1. For this, a ray strikes the surface at an angle Φ and must be reflected at the same angle. This ray must pass through the focus, f, by definition. Then
$r \cot 2\Phi + y = f$
and the distance $A = H - y(r)$ and $B^2 = (f - y(r))^2 + r^2$
So the path from the pupil plane to the focus is $A + B$.
(b) From Fig. 6.11, $2\alpha + 2\Phi = 180°$. In terms of the value in Fig. 6.11, $\Phi = 180° - \alpha$, and $\alpha = dy/dr$. In addition, the trig identity gives:
$\cot 2\Phi = 1/2\left((1 - \tan \Phi^2)/\tan \Phi\right)$
The focus is $f = r \cot 2\Phi + y = r/2\left(1 - (dy/dr)^2\right)/(dy/dr) + y$
For a surface $y = r^2/4f$, the focus is a constant, as is $A + B$.
(c) No, you need an additional reflector. This is shown by the fact that the result of a reflection from the surface of a sphere must form an isosceles triangle with a hypotenuse of length equal to the radius of the sphere. Then the distance from the point of the reflection to the focus is one half of the radius, R, times the cosine of the angle of reflection. The distance to the center of the sphere from the lowest point is the sum of the focus plus $R/2\times \cos \Phi$. Then the focal point, f, is $f = R/2 \times (1 - 0.5/\cos \Phi)$. For very small angles, there is a unique focus, but not for larger angles.

9. Eq. (6.56) from 'Tools' is:

$$dE_y(\phi) = -\mathrm{i}\, J_0\, g(x')\, \frac{1}{r} \mathrm{e}^{-\mathrm{i}\,(\omega t - k r)}\, dx' \ .$$

This is a one-dimensional expression. For Eq. (6.57), we have:

$$dE_y = -\frac{\mathrm{i}}{2} \lambda J_0\, g(\mathbf{x}')\, \frac{F_e(\mathbf{n})}{|\mathbf{x} - \mathbf{x}'|} \mathrm{e}^{-\mathrm{i}\,(\omega t - k\,|\mathbf{x} - \mathbf{x}'|)}\, \frac{dx'}{\lambda}\frac{dy'}{\lambda}$$

using $|\mathbf{x} - \mathbf{x}'| = r$ and $\frac{1}{2}\lambda F_e = 1$, then Eq. (6.57) is directly related to Eq. (6.56).

10. Place two isotropic emitters along the x axis. Assume these emit put sine waves. Label these no. 1 and no. 2. Place no. 1 to the left of no. 2. These are separated by $\lambda/4$ The emission from both has the same amplitude. If the phase of the emission from no. 1 lags $\lambda/4$ behind that of no. 2, the emission from no. 1 is $\lambda/2$ at no. 2, so cancels along the direction of the x axis to the right. To the left, these reinforce. In other directions, the result is intermediate.

11. From problem 8 of chapter 3, or alternately, the theory of images, there is an identical emitter, with the same phase, at a depth of $\lambda/4$ to cancel the incoming radiation which is perpendicular to the surface. In other directions, the cancellation is lower. The distance between the emitters is double that in problem 10, so the angle at which the cancellation occurs will differ.

12. Eq. (6.78) in 'Tools' is

$$\mathrm{FWHP} = 1.02\,\frac{\lambda}{D}\,\mathrm{rad} \simeq 58.4°\,\frac{\lambda}{D}$$

Convert degrees to arc seconds using the factor 3600. To use D in meters and λ in millimeters requires a factor of $1/1000$. So the terms are $(3600/1000) \times 58.4 \times 1.02 = 214.4$. These units are appropriate for single dish millimeter telescopes such as the IRAM 30-m or the 50-m Large Millimeter Telescope (LMT) in Mexico. This is appropriate for a fully illuminated dish; usually there is some taper to reduce sidelobes. Compare to figure 22.1.

13. Eq. (6.41) is: $P_S = \frac{c}{3}\left(\frac{I\Delta l}{2\lambda}\right)^2$ If we require that

$$\frac{I\Delta l}{2\lambda} = \frac{e\dot{V}}{c^2}$$

we have

$$P_S = \frac{c}{3}\left(\frac{e^2 \dot{V}^2}{c^4}\right)^2$$

If

$$e^2 \dot{V}^2 = 64\pi^4 v^4 \mu^2$$

This is the quantum mechanical expression for spontaneous emission:

$$A_{mn} = \frac{64\pi^4}{3\,h\,c^3}\, v_{mn}^3\, |\mu_{mn}|^2$$

given in Eq. (12.24) of 'Tools' and many other texts. In this expression, $|\mu_{mn}|$ is the dipole moment which has units of charge times distance.

14. Eq. (6.32) is: $A_z = \frac{1}{c} \frac{I \Delta l}{r} e^{-i(\omega t - k r)}$ inserting $I(z) = I_0 \cdot \left(1 - \frac{|z|}{\Delta l/2}\right)$ into

Eq. (6.40): $|\langle S \rangle| = \frac{c}{4\pi} |\operatorname{Re}(\mathbf{E} \times \mathbf{H}^*)| = \frac{c}{4\pi} \left(\frac{I \Delta l}{2\lambda}\right)^2 \frac{\sin^2 \vartheta}{r^2}$ The calculation of the average current $I(z)$ is

$$\hat{I} = \frac{\int_0^{\Delta l/2} (I(z))\, \mathrm{d}z}{(\Delta 2)}$$

The value of this average is $I/2$. Inserting this into Eq. (6.41), we obtain a radiated power of

$$P_S = \frac{c}{3} \left(\frac{I \Delta l}{4\lambda}\right)^2$$

Since the current is squared, this shows that $1/4$ of the power radiated if the current is a constant. The equation defining the Radiation Resistance is:

$$[P] = \tfrac{1}{2} R I^2$$

since the power, P, is $1/4$, the value of R is:

$$R_S = \frac{c}{24} \left(\frac{\Delta l}{\lambda}\right)^2$$

Chapter 23
Solutions for Chapter 7: Practical Aspects of Filled Aperture Antennas

1. (a) $I_\nu = 2kT_B/\lambda^2$ and $S_\nu = \int (2kT_B/\lambda^2)\mathrm{d}\Omega = (2kT_B/\lambda^2)\Delta\Omega$.

We have $W = \frac{1}{2} A_e (I_\nu \Omega_A) \Delta\nu = A_e (S_\nu) \Delta\nu$.

Since $W = kT_A \Delta\nu = \frac{1}{2}A_e S_\nu \Delta\nu$, we have

$T_A = \frac{1}{2}k A_e S_\nu$.

For brightness temperature, we need the relation of antenna beam and source size; use Eq. (7.23) in 'Tools'

$T_A = \left(\int T_B P_n \mathrm{d}\Omega\right) / \left(\int P_n \mathrm{d}\Omega\right)$.

If the source is small, the integral in the numerator is $\int T_B P_n \mathrm{d}\Omega \approx T_B P_n(0) \Omega_{MB} = T_B \Omega_{MB}$, while the denominator is $\left(\int P_n \mathrm{d}\Omega\right) = \Omega_A$.

Since $\Omega_{MB}/\Omega_A = \eta_B$, we have

$T_A = \eta_B T$

(b) As measured with a telescope beam large compared to the source, we have:

$S_\nu = 2k \int (T/\lambda^2) \mathrm{d}\Omega = (2kT_B/\lambda^2) \left(1.133 \left[\theta_s^2 + \theta_B^2\right]\right)$

For a source very large compared to the telescope beam

$S_\nu = (2kT/\lambda) \mathrm{d}\Omega = (2kT_0/\lambda) \left(1.133 \left[\theta_s^2\right]\right)$.

Setting these last two equations equal, we have

$T_0 = T_B \left[\theta_s^2 + \theta_B^2\right] / \theta_s^2$.

2. $T_A = \eta_B T_0 \, \Omega_s/\Omega_{MB} = \eta_B T_0 (\theta_s/\theta_B)^2 = (0.6) \times (600\,\mathrm{K}) \times (0.667/8)^2 = 2.5\,\mathrm{K}$.

3. Use the symmetry about the center of the beam: $P = 2I \int_0^{2\pi} \int_0^{10°} \sin\theta \, \mathrm{d}\theta = 4\pi I \left(1 - \cos(10°)\right) = 0.19\,I$. So $S = I/(0.19) \times (60 \times 10^3\,\mathrm{m})^2) = 1.45 \times 10^{-4}\,\mathrm{W\,m^{-2}}$. To receive $10^{-6}\,\mathrm{W}$, we need $7 \times 10^{-2}\,\mathrm{m^2}$, or an antenna with dimensions of 8 cm on a side. The actual antenna must be larger, since there will be losses and the efficiency will be less than 100%.

4. $\Omega_A = P_{max} \int_0^{2\pi} \mathrm{d}\phi \int_0^{1°} \sin\theta \, \mathrm{d}\theta + \int_{1°}^{10°} \sin\theta \, \mathrm{d}\theta + 0$

From this, we have

© Springer International Publishing AG, part of Springer Nature 2018

T. L. Wilson, S. Hüttemeister, *Tools of Radio Astronomy – Problems and Solutions*, Astronomy and Astrophysics Library, https://doi.org/10.1007/978-3-319-90820-5_23

$2\pi\,(1-\cos\,(1°)) + 2\pi(\cos\,(1°) - \cos\,(10°)) = 2\pi(1.5 \times 10^{-4} + 1.5 \times 10^{-3}) = 2\pi(1.52 \times 10^{-3})$.

$\Omega_{MB} = P_{max} \int_0^{2\pi} d\phi \int_0^{1°} \sin\theta\,d\theta = 2\pi\,(1.5 \times 10^{-4})$.

$\eta = \Omega_{MB}/\Omega_A = 0.1$.

5. Yes, but the antenna need not be special. From Problem 6, since $T_A = T_0\,\Omega_s/\Omega_A = \eta_B T_0\Omega_s/\Omega_{MB}$. We assume that $\Omega_s < \Omega_{MB}$, so we always have $T_A < T_0$.

6. No, since $\theta^2_{observed} = \theta^2_{beam} + \theta^2_{source}$. So the observed size is *always* larger than the source size.

7. Set $x^2/2\sigma^2 = 4\ln 2x^2/\theta^2_{1/2}$. Solving for $\theta_{1/2}$ we obtain $\theta_{1/2} = \sqrt{8\ln 2}\sigma = 2.355\sigma$

8. (a) Make the ring flat:

$A_{geom} = 2\pi \int r\,dr = \pi\,(R^2_{outer} - R^2_{inner}) = \pi\,((305 + 15)^2 - (305)^2) = 2.95\;10^4\,m^2$

(b) Use Eq. (6.70), which is:

$$P_n(u) = \left[\frac{\displaystyle\int_0^\infty g(\rho)\,J_0(2\pi u\rho)\rho\,d\rho}{\displaystyle\int_0^\infty g(\rho)\rho\,d\rho}\right]^2$$

with a grading $g(\rho)=1$ for $R_{in} < R < R_{outer}$. Since the R's are nearly equal, one can extract the Bessel function from the integration. Then using a value $\bar{R}=1/2(R_{in} + R_{outer})$ for the argument of the Bessel function, we have a simpler result.

We call the integral over the grading D, where $D=\int \rho d\rho = 1/2\,(R^2_{outer} - R^2_{inner})$

The far field pattern is then $P_n=(N/D)^2$, with

$N = D\,(J_0(2\pi \sin\,(\theta\bar{R}/\lambda)))^2$

For small angles, $\sin\,(\theta\bar{R}/\lambda) = \theta\bar{R}/\lambda$

$P_n = \frac{1}{D}\,(J_0(2\pi\theta\bar{R}/\lambda))^2$

The amplitude of the first sidelobe is $(0.4)^2 = 0.16$. This is at $x = 3.9 = 2\pi\theta\,(\bar{R}/\lambda)$, or $\theta = 1.24\,(\lambda/\bar{R})$

The full width to half power of the main lobe is at $x = 1.5$, so using the relations above, have $x=1.5=2\pi\theta\,(\bar{R}/\lambda)$, or $\theta = 0.95\,(\lambda/\bar{R})$

(c) From the result of problem 1 of chapter 7, have

$T_A = \eta_A S_\nu\,A_g/2k = 0.5\cdot S_\nu(2.94\;10^4\,m^2)/2(1.38\cdot10^{-23}) = 5.33\,S_\nu$.

9. You need a 50/1 signal-to-noise ratio. There is no contradiction. A wide beam can still be pointed very accurately in a given direction, given a high signal-to-noise ratio.

10. The surface accuracy must be 0.02λ, from Fig. 7.7 in 'Tools', which is given in the statement of this problem.

(a) The value of $\eta_A = 0.72$.

(b) From Fig. 7.7, need $K/(K+1)=1$, so $\eta_A = 0.94$. The beam efficiency is 0.82. From Table 6.1 of 'Tools', A frequency of 28 GHz is equivalent to a wavelength of 1.06 cm. Then $\theta = 1.02\,(\lambda/D)=1.02\,(1.06 \cdot 10^{-2}/7) = 1.55 \cdot 10^{-3}\,\mathrm{rad}=5.3'$. See figure 22.1 for the far-field antenna power pattern.

11. Eq. (6.52) gives a definition of the quantity $\theta_{geom} = \lambda/D$, where θ_B is the measured main beam Full Width to Half Power. In addition, we must use Eq. (7.5). This is $\eta_B = \Omega_{MB}/\Omega_A$ In addition, we need Eq. (7.9). This is $A_e = \eta_A \times A_{geom}$ We also need Eq. (7.11). This is: $A_e \times \Omega_A = \lambda^2$ Combining all of these relations, we have:

$$\eta_B = \Omega_{MB} \times A_e/\lambda^2 = \eta_A\,\Omega_{MB}A_{geom}/\lambda^2 = 1.133 \times \eta_A\,\theta_B^2\pi/4\,D^2/\lambda^2$$

This gives

$$\eta_B = 1.133\,\theta_B^2\,\eta_A\,\pi/4\,D^2/\lambda^2 = 1.133\eta_A\,\pi/4\,\theta_B^2/\theta_{geom}^2.$$

12. As before, $\theta_{geom} = \frac{\lambda}{D}$; we need Eq. (7.9) from problem 11, and the definition of SEFD, Eq. (7.27) of 'Tools': $\text{SEFD} = \frac{2\,\eta'\,k\,T_{sys}}{A_{eff}}$. Then, from these definitions:

$$\text{SEFD} = \frac{2\,\eta'\,k\,T_{sys}}{\eta_A\pi/4D^2} = \frac{2\,\eta'\,k\,T_{sys}}{\eta_A\pi/4D^2} = \frac{8\eta'\,k\,T_{sys}\times\theta_{geom}}{\pi\,\eta_A\,\lambda^2}$$

13. Eq. (6.40) is given in the solution of problem 14, chapter 6: This equation describes the total radiated power. The angular distribution is $P = P_0 \sin\theta^2$ and the normalized power pattern is $P_n = \frac{P}{P_0}$ An important quantity is the beam solid angle, Ω_A This is defined as in Eq. (7.3), given in the statement of this problem. From the evaluation of this, the value is $\Omega_A = 8\pi/3$. From Eq. (7.5) of 'Tools', the main beam efficiency can be obtained.

Using this definition, the result is $4\pi/3$. Then making use of $\eta_B = \frac{\Omega_{MB}}{\Omega_A}$ This is

$$\eta_B = \frac{4\pi/3}{8\pi/3} = 1/2$$

From Eq. (7.11), $A_e \times \Omega_A = \lambda^2$ The value of Ω_A is $4\pi/3$, so $A_e = 3\lambda^2/4\pi$

14. The power from the tapered Hertz dipole is 1/4 of that from a Hertz dipole that has a uniform current, since the integral along the dipole is

$$\int_0^{\Delta l/2} I_0 \times \left(1 - \frac{|z|}{\Delta l/2}\right) dz$$

The result of this integral is $\frac{1}{2}I_0\left(\frac{\Delta l}{2}\right)$ Since the power depends on the value of I_0^2, this is 1/4 of the power emitted by a Hertz dipole with a constant current. The direction of maximum power is independent of angles, so maximum gain will be 1/4 of that from a Hertz dipole with a uniform current (see problem 14 of Chapter 6). Thus, the value of G_{max} is the same as for the dipole with a constant current. From Eq. (7.6):

$$G_{max} = \frac{4\pi \, A_e}{\lambda^2}$$

Since Ω_A depends only on angles, not on the distribution of current, Ω_A is the same as for the dipole with a uniform current, so the result is $8\pi/3$. From Eq. (7.11), $A_e \, \Omega_A = \lambda^2$

Thus, the effective area of this dipole antenna, A, must be the same as the uniform current case.

15. This is the first term in Taylor series expansion of the difference between a wave expanding from a point source (i.e. an expanding spherical wave) and a plane wave. Putting the dimensions in meters:

$$k = \frac{2d^2}{\lambda} = \frac{2 \times (100)^2}{0.03} = 6.7 \times 10^5 \, m = 670 \, km$$

Thus the radiation is a plane wave at this distance. For the second example, for the dimensions in meters

$$k = \frac{2 \times (10)^2}{0.007} = 29 \, km$$

16. Eq. (7.26) is $S_\nu = 3520 \frac{T_A'[K]}{\eta_A[D/m]^2}$, where T_A' is the antenna temperature corrected for atmospheric attenuation. In most cases, in the centimeter wavelength range, this correction is small. So, then: $S_\nu = 3520 \frac{T_A'}{(0.5) \times (305)^2} = 7.6 \times 10^{-2} \, T_A'$

For a 100 m aperture, the result is: $S_\nu = 3520 \left(\frac{T_A'}{(0.5) \times (100)^2} \right) = 0.70 \, T_A'$.

Chapter 24
Solutions for Chapter 8: Single Dish Observational Methods

1.(a) The noise from the atmosphere is $250\,(1 - e^{-0.1}) = 23.7\,$K.
The source intensity is reduced by a factor $(e^{-0.1}) = 0.90$, so 10% of the source intensity is lost.
(b) The results, obtained in the same way as in part **(a)**, are given in Table 24.1.
(c) $\tau(30°) = \tau_z / \sin 30° = 2\,\tau_z$
$\tau(20°) = \tau_z / \sin 20° = 2.92\,\tau_z$
(d) $\tau(15°) = \tau_z / \sin 15° = 3.86\,\tau_z$
$\tau(10°) = \tau_z / \sin 10° = 5.76\,\tau_z$
There is a 32% increase in the absorption between 20° and 15°, and a 200% increase between 20° and 10°.
(e) The emission from the atmosphere will raise the system noise. For $\tau = 0.2$, have an extra 36.2 K. This increases the receiver noise by 36%. However, one must also correct the astronomical intensity for the absorption. This is a 22% effect. Overall, both effects, absorption of the signal and emission from the atmosphere, must be accounted for. This results in a much greater worsening of the receiver sensitivity. For the system 100 K receiver noise, the effective system noise is 166 K, or a 66% worsening of the sensitivity. For a receiver with a noise temperature of 20 K, have the same increase from atmospheric emission, but the increase caused by the loss in signal is less, so the final system noise is 69 K. However the increase is 343%.

2. The noise from the atmosphere is $200\,(1 - e^{-\tau})$. At 30°, the τ is twice that at 90°, so at this elevation, the τ is
$T = 200\,(1 - e^{-\tau})$, or $\tau = 0.097$. At 30°, get $\tau = 0.194$, which gives $T = 35.2\,$K, so the data are consistent. At 60°, get 21.1 K, 20°, get 49.4 K, so the temperature of the atmosphere is perhaps 5% too large. Using a least squares fit, one could solve for both τ and the temperature of the atmosphere assuming that a plane parallel model is adequate (Table 24.2).

© Springer International Publishing AG, part of Springer Nature 2018
T. L. Wilson, S. Hüttemeister, *Tools of Radio Astronomy – Problems and Solutions*, Astronomy and Astrophysics Library,
https://doi.org/10.1007/978-3-319-90820-5_24

Table 24.1 Data for atmospheric optical depth, emission and transmission as treated in problem 1(b)

Optical depth	Emission from atmosphere (K)	Percent transmission of signal (%)
0.5	118	61
0.7	151	50
1.0	190	37
1.5	233	22

Table 24.2 Atmospheric optical depth relative to 225 GHz; results for Problem 2

Frequency ν (GHz)	Optical depth ratio $\tau(\nu)/\tau$ (225 GHz)
345	3.4 ± 0.5
460	11.2 ± 3.0
807	18 ± 7

3. Without snow in the dish, the sky radiates 30 K. Take the fraction of the dish covered with snow as f. Then
$f \times (270) + (1 - f) \times (30) = 130$. Solving for f, have $f = 0.42$.

4. At the maximum elevation, $\tau = 0.1/\sin 11° = 0.524$
At the minimum elevation, $\tau = 0.1/\sin 8° = 0.718$. The ratio is 1.37, so there is a 37% increase in τ. So the comment is reasonable. The emission from our atmosphere is 102.5 K at $8°$ and 81.5 K at $11°$ elevation. The reduction in the intensity of the astronomical signal is
0.592 at $11°$ elevation, and
0.487 at $8°$ elevation.
At $8°$ elevation, $\Delta T_{RMS} = 142.5/\sqrt{40 \times 10^6} = 0.022$ K
At $11°$ elevation, $\Delta T_{RMS} = 121.5/\sqrt{40 \times 10^6} = 0.019$ K

5. For the solid angle, the relation to a Gaussian beam is $\Delta\Omega = (1.06\,\theta_{1/2})^2$. Where $\theta_{1/2}$ is in radians. Converting from radians to arcmin, we have $\Delta\Omega = 8.46 \times 10^{-8}\,(\theta_{1/2})^2$. Converting from meters to cm in the expression for λ, we use k=1.38 $\times 10^{-23}$ W K^{-1}, and note that Jy is 10^{-26} in MKS units (10^{-26} W m^{-2} Hz^{-1}). Then we have:
$$S(Jy) = 2 \times (1.38 \times 10^{-23})T(8.46 \times 10^{-8}\theta_{1/2})^2/(10^4\,\lambda(cm)^2$$

The result is Eq. (8.19) in 'Tools': $S(Jy) = 2.65\,T \times \left((\theta_{1/2})^2/(\lambda^2)\right)$
where θ is in units of arc minutes, and λ in centimeters.

6. (a) Taking the expression for flux density from the last problem, or from problem 10 of chapter 1, we have:

$$T(\text{actual brightness}) \times (\theta(\text{actual angular size}))^2 = T(\text{observed}) \times (\theta(\text{observed}))^2$$

where the observed temperature is the Main Beam Brightness Temperature. If we set T(actual brightness)=T(source) and T(observed)=T(Main Beam) and θ(actual size) $= \theta$(source), so we find that:

$$T(source) = T(main\ beam) \times \left(\frac{\theta(observed)}{\theta(source)} \right)^2$$

Since θ(observed) $> \theta$(source), we must have T(source)>T(main beam)
For Gaussian source and beam shapes, from Eq. (8.47) or the results given in problem 1 of chapter 7, we have

$$T_A(x, y) = \frac{\int P(x - x', y - y')\, T_B(x', y')\, dx'\, dy'}{\int P(x', y')\, dx'\, dy'}$$

We will use Gaussians in the integrals, and restrict the integration to 1 dimension. Then, we have: $T_A(x) = \left(\int e^{-(x-x')^2/\sigma_1^2} T_0 e^{-x'^2/\sigma_2^2}\, dx' \right) / \left(\int P(x')\, dx' \right)$.
We use the "brute force" first, and consider only the numerator, with the proper normalization:
$u(x_0)=1/(2\pi\sigma_1\sigma_2) \int_{-\infty}^{+\infty} e^{-(x-x_0)^2/2\sigma_1^2} e^{-x^2/2\sigma_2^2}\, dx$.
In the following, we concentrate on the exponential factors. Combining the two, we have, $(x_0^2 - 2x\,x_0 + x_0^2)/2\sigma_1^2 + 1/2\sigma_2 x^2 = x^2(1/2\sigma_1^2 + 1/2\sigma_2^2) - x\,x_0/\sigma_1^2 + x_0^2/2\sigma_1^2$.
We introduce an expression $\alpha = (1/2\sigma_1^2 + 1/2\sigma_2^2)$, obtaining:
$\alpha x^2 - x\,x_0/\sigma_1^2 + x_0^2/2\sigma_1^2 = \alpha \left(x - x_0/2\,(\sigma_1^2\alpha) \right)^2 - x_0^2/(4\,\sigma_1^4\alpha) + x_0^2/(2\sigma_1^2)$.
The last two terms do not involve the variable of integration, so can be factored outside the integral:
$\frac{x_0^2}{2\sigma_1^2} \left(1 - \frac{1}{2\sigma_1^2\alpha} \right) = \left(\frac{x_0^2}{2(\sigma_1^2 + \sigma_2^2)} \right)$.
Then we have:
$u(x_0) = \frac{1}{2\pi\sigma_1\sigma_2} e^{-(x_0^2/2(\sigma_1^2+\sigma_2^2)} \int_{-\infty}^{+\infty} e^{-\alpha(x-x_0/2\sigma_1^2\alpha)^2}\, dx$.
The integral gives
$u(x_0) = \frac{\sqrt{2\pi}}{2\alpha\pi\sigma_1\sigma_2} e^{-x_0^2/(2[\sigma_1^2+\sigma_2^2])} = \frac{\sqrt{2\pi}}{\alpha\,\sigma_1\sigma_2} e^{-x_0^2/(2(\sigma_1^2+\sigma_2^2))}$
The relation between the total σ and the two individual σ terms is
$\sigma^2 = \sigma_1^2 + \sigma_2^2$
The shorter method involves Fourier Transform (FT) relations. Use Table 3 from Appendix A of 'Tools'. The convolution in x is the product of Fourier Transforms in s space and
$FT \left(e^{-x^2/2\sigma_1^2} \right) \leftrightarrow e^{-2\pi\sigma_1^2 s^2}$.

The expression for σ_2 is similar. Then, we have $u = y \otimes x \leftrightarrow e^{-2\pi(\sigma_1^2+\sigma_2^2)\,s^2} = e^{-2\pi(\sigma^2)\,s^2}$. Transforming back to x space, we have $FT \left(e^{-2\pi(\sigma^2)\,s^2} \right) \leftrightarrow e^{-x^2/2\sigma^2}$.
Then $\sigma^2 = \sigma_1^2 + \sigma_2^2$, as before. For the application in question, $\theta_o^2 = \theta_s^2 + \theta_B^2$.

7. $P = S(\text{Jy}) \times A_e \times \Delta\nu/2$ for a point source.
$A_e = 3120\,\text{m}^2$, so
$P = 10^{-20}\,\text{W}\,\text{m}^{-2}\,\text{Hz}^{-1} \times 3120\,\text{m}^2 \times 10^6\,\text{Hz}/2 = 1.56 \times 10^{-10}\,\text{W}$.
Use the relation that (received maser power/m^2/(Dangerous Power Level/m^2) $\times$ (source distance)^2 is the square of the dangerous distance. D is the danger distance, so
$D = \sqrt{10^{-24}/10^{-2}}\,1.5 \times 10^{21}\,\text{cm} = 1.5 \times 10^{10}\,\text{cm}$.
This is not very realistic, since H_2O masers have sizes of about 10^{13} cm, so the 'point source' assumption breaks down. However the example shows that astronomical sources emit little power.

8. For Sun, take $\theta_{max} = 15'$, with $T = 5800\,\text{K}$, so
$S = \frac{2kT}{\lambda^2} \int \sin\theta\,d\theta \int_0^{2\pi} d\phi$
$S = \frac{4\pi kT}{\lambda^2}(1 - \cos\theta_{max})$
where $\theta_{max} = 0.9999$, so $S = (1.01 \times 10^{-18}\,\text{W}\,\text{Hz}^{-1})\,(10^4\,\text{m}^{-2})(9.52 \times 10^{-6}) = 9.57 \times 10^{-20}\,\text{W}\,\text{m}^{-2}\,\text{Hz}^{-1} = 9.57 \times 10^6$ Jy.

9. (a) $S_\nu = 5.4\,\text{Jy} = 2.65\,T_{MB}\,(43''/60'')^2/(1.3\,\text{cm})^2 = 2.65\,(0.3039)\,T_{MB}$, so $T_{MB} = 6.7\,\text{K}$. This is the Main Beam brightness temperature.

(b) The source has an actual temperature of $T_s = T_{MB}\left(\frac{43}{10}\right)^2$, so $T_s = 124\,\text{K}$.
Set $T_{MB} = 124\,\text{K} = 14\,000(1 - e^{-\tau})$, so $\tau = 0.009$.

10. The wavelength is 3 m, or 300 cm. Then
$1\,\text{Jy} = 2.65 T_s\left[(10/60)^2/(300\,\text{cm})\right]^2 = 8.18 \times 10^{-7}\,T_s$.
The temperature is $T_s = 1.22 \times 10^6\,\text{K}$.
The observing wavelength is 30 cm. The source temperature is 100 times less. $T_s = 1.22 \times 10^4\,\text{K}$.

11. Use the definition of an Astronomical Unit, so $D = (0.227) \times (1.46 \times 10^{13}) = 4.04 \times 10^{12}\,\text{cm}$
The FWHP angle is:
$$\theta = 2\frac{6100 \times 10^5\,\text{cm}}{4.04 \times 10^{12}\,\text{cm}} = 1.51 \times 10^{-4}\,\text{rad} = 62.2''$$

Then $T_{\text{Venus}} = 8.5 \times \left(\frac{8.7 \times 60''}{62.2''}\right)^2$,
so $T_{\text{Venus}} = 600\,\text{K}$. From a further analysis, it has been determined that the high surface temperature of Venus is caused by the 'greenhouse effect'.

12. This is an incomplete statement: the sky noise which dominates at or below 400 MHz is source or astronomical sky noise, which one wants to measure. The sky noise in the sub-millimeter wavelength range is from the earth's atmosphere. Thus in the sub-millimeter wavelength range, sky noise has only a negative effect. As shown in Problem 6 of this chapter, the sky absorbs signals and radiates, raising the system noise. The latter effect is multiplicative, so a low system noise helps to improve the signal-to-noise ratio in the measurements.

13. (a) The transmission is equal to $e^\tau = e^{-3} = 0.00498$.

(b) $T_{sky} = 200(1 - e^{-3}) = 200(0.95) = 190$ K.

(c) The system noise in the case of the 'corrected antenna temperature', that is, the system noise outside the earth's atmosphere and corrected for antenna efficiency, is $(50\,K + 190\,K)/(0.0498) = 4820$ K.

(d) For a 12 m telescope, the antenna temperature of a small diameter source is $T_A^* = 0.5(0.0196)S_v(Jy) = 9.810^{-4}$ K. $\Delta T_{RMS} = \frac{2 \times 4820}{\sqrt{2\,10^9 \times t}} = 0.216/\sqrt{t}$, so $t = 13.5$ h.

It is better to use a bolometer.

14. Solving for v/c yields $v/c = [(1 + z)^2 - 1]/[(1 + z)^2 + 1]$; then for $z = 2.28$, $v/c = 0.830$, for $z = 5$, $v/c = 0.946$ and for $z = 1000$, $v/c = 0.999998$.

15. The effect is the correlation of incident and reflected voltages, V_i and V_r. Thus $V_i \cdot V_r = 10^{-3}$. Taking $V_i = 1$, have a power flux in the reflection of $(V_r)^2/377\,W\,m^{-2}$. As a fraction of the incident power, must use the ratio of the squares of the voltages. This is 10^{-6} of the incident power.

16. The average number of $1'$ sources per unit area has the average value, P, $P = \lambda = 125 \frac{1' \times 1'}{15' \times 120'} = 6.94 \times 10^{-2}$. The assumption is that the sources do not overlap. Thus about 6% of the cloud is filled with dense gas, so 94% is filled with low density gas.

$P(Poisson) = m^n\,e^{-m}/x!$,

where x is the number of sources expected. For $x = 1$, have P=0.0647, for $x = 2$, P=0.00225, for $x = 3$, P=5.2 × 10^{-5}, for $x = 4$, then P=9.02 × 10^{-7}.

17. (a) Take $\lambda = 0.04$, then for two sources in the same beam, P=7.7 × 10^{-4}, for three sources, P(3)=10^{-5}

(b) If take 10^6 samples, have 780 cases where sources are confused. This is to be compared to a total of 40,000 sources. So this is an unconfused survey.

18. Start with $dn \propto S^{-\gamma}\,dS$. Then have

$$\bar{n} = \int n(S)\,d\Omega\,dS = k \int (fS)^{-\gamma} f^\gamma\,d\Omega \frac{d(fS)}{f}$$

Use $x = fS$, so that

$$\bar{n} = k \int\int x^{-\gamma}(f^{\gamma-1})\,d\Omega\,dx = k\Omega_e \int x^{-\gamma}\,dx$$

from the relation $\Omega_e = \int f^{\gamma-1}\,d\Omega$ then, $\bar{n} = k\Omega_e \int x^{-\gamma}\,dx$ We need to calculate σ^2. This is given by:

$$\sigma^2 = \int_0^{D_c} x^2\,d\bar{n} = \int_0^{D_c} k\,x^{-\gamma}\Omega_e\,dx$$

This is: $\sigma^2 = k \int_0^{D_c} \Omega_e \, x^{-\gamma+2} \, dx$ which gives $\sigma^2 = k \, \Omega_e \left(\frac{1}{3-\gamma}\right) \times D_c^{3-\gamma}$ Since $D_c^{3-\gamma} = (q \, \sigma)^{3-\gamma}$, then the final result is

$$\sigma^2 = \left(\frac{k \, \Omega_e \, q}{3 - \gamma}\right)^{1/(\gamma-1)}$$

The term $k \, \Omega_e$ can be factored out of the brackets to obtain the final result.

19. (a) Inserting the expression for a Gaussian beam in Eq. (8.59), we have, in 1 dimension (x), and with $4 \ln 2 = 2.77$:

$$\Omega_e = \int \left[A e^{2.77 \, x^2/\theta_{1/2}^2}\right]^{\gamma-1} dx$$

where we have used the symbol x for the angle. Using

$$u = \sqrt{\frac{2.77 \, (\gamma - 1)}{\theta_{1/2}}} \times x$$

and noting that the integral of a Gaussian is $\sqrt{\pi}$, we have

$$\Omega_e = \left(\sqrt{\pi} \, A^{\gamma-1} \, \theta_{1/2}\right) / \left(\sqrt{4 \ln 2 \, (\gamma - 1)}\right)$$

(b) From 'Tools', Eq. (8.58), we have $d\bar{n} = k \, \Omega \int x^{-\gamma} \, dx$, where is the number of sources with flux densities between S and $S + dS$. Solving for k, the result is

$$k = \frac{\bar{n}}{\Omega_e \, x^{3-2\gamma} \, (1 - \gamma)^2}.$$

From the statement of the problem, $k = \gamma \, N_c \, S_c^{-\gamma} = 1.5 \times 10^5 \times (10^{-2})^{-1.5}$, where the units are per steradians $\times$ Janskys.

20. Assume that $S = \frac{P_0}{4\pi R_0^2}$ or $R_0^3 = (P_0)^{1.5} / \left((4\pi)^{1.5} \, S_0^{1.5}\right)$
$S > S_0$ means $R < R_0$, since all sources are intrinsically the same. Imagine the universe as a sphere with a density of sources ρ. The total number of sources with a flux density larger than a certain value S_0 is

$$N(S > S_0) = \frac{4\pi R_0^3 \rho}{3} = \frac{\sqrt{4\pi} R_0^3 \rho}{3 S_0^{1.5}} = K S_0^{-1.5}.$$

Where $K = \frac{\sqrt{12\pi}}{R_0^3 \rho}$ The number of sources varies inversely as the 1.5 power of flux density, and the differential number varies inversely as the 2.5 power of flux density. This is a 'Euclidean universe'.

21. The probability density for the number of sources per steradian with a flux between S and $S + dS$ is $p(S)$ that is, on the average we will observe $\bar{v} = \Omega\, p(S)\, dS$ sources with flux densities in the interval $(S, S + dS)$ per beam Ω. The average number of sources per beam is $\bar{n}$. If the sources are distributed according to a Poisson distribution (see "An Introduction to Error Analysis", J. R. Taylor, University Science Books) the probability of n sources in a beam is $f(n) = \frac{\bar{v}^n}{n!} e^{-\bar{v}}$ The first moment is $\mu_1 = \bar{v}$. The second moment, μ_2, is $\mu_2 = \sum_{v=0}^{\infty} v^2 f(v) = \bar{v}(\bar{v}+1)$ Then the dispersion is $\mu_2 - \mu_1\, \sigma_v^2 = \bar{v}$ When applied to flux densities, this is $\sigma_S^2 = S^2\, \bar{v}$.

(b) The shape of the beam is a "pill-box" with perpendicular walls, so that the total output of the antenna is the unweighted sum of the flux densities of all sources within the beam and zero for sources outside. $S = \sum_k S_k$. The average $\bar{S}$ caused by sources $(S, S + dS)$ will be $\bar{S} = S\,\bar{v}$.

The total signal is caused by sources with different flux densities, so the dispersions add quadratically. Those sources with flux densities between S_c and S_L and a number density of $p(S)$ then result in a dispersion for the total flux of $\sigma_{S_c}^2 = \Omega \int_{S_L}^{S_c} S^2\, p(S)\, dS$ Then sources with a flux density cut-off S_c can be measured with a signal to noise ratio, q, of $q = \frac{S_c}{\sigma_{S_c}}$ where S_L is a lower limit to the actual flux density, where S_c is a cutoff. Rearranging terms, we have: $q^2 = \frac{S_c^2}{\Omega} \frac{1}{\int_{S_L}^{S_c} S^2\, p(S)\, dS}$ The distribution of faint point sources is $p(S) = n\, N_c\, S_c{}^n\, S^{-n-1}$ for a wide range of S, the observed value of n is 1.5, so:

$$q^2 = \left(0.333 \frac{1}{\Omega N_c} \frac{1}{1 - (S_L/S_c)^{0.5}}\right);$$

(c) For the case $S_L = 0$, we have:

$$\lim_{S_L \to 0} q^2 = 0.333 \frac{1}{\Omega N_c},$$

Taking $S_L = 0.666\, S_c = 1.0$, q is taken to be 5, and ΩN_c is interpreted as the number of sources per beam area, we have: $\Omega\, N_c = 7.7 \times 10^{-3}$. This implies very few sources per beam area.

Chapter 25
Solutions for Chapter 9: Interferometers and Aperture Synthesis

1. Use coordinates x and u. Then $I'(x) = \int V(u)e^{-2\pi i u x} du$ for all parts of this problem.

(a) $I'(x) = \int_0^{u_0} V(u)e^{-2\pi i u x} du$. So
$I'(x) = (1/2\pi i x)\left(1 - e^{-2\pi i u_0 x}\right) = 2u_0 e^{-2\pi i u_0 x}\left(\sin\left(2\pi u_0 x\right)/2\pi u_0 x\right)$.
For large u_0, the zero crossing is at small x values. These negative sidelobes will distort the image in the x coordinate.

(b) $I'(x) = \int_{u_{\min}}^{u_{\max}} V(u)e^{-2\pi i u x} du$
$= \int_0^{u_{\max}} V(u)e^{-2\pi i u x} du - \int_0^{u_{\min}} V(u)e^{-2\pi i u x} du$

$= \int_0^{u_{\max}} V(u)e^{-2\pi i u x} du - \left[2u_0 e^{-2\pi i u_0 x}\left(\sin\left(2\pi u_0 x\right)/2\pi u_0 x\right)\right]$.
The second term gives rise to a 'negative bowl' in the x coordinate, which is the image plane.

2. Eq. (9.6) is

$$R(\mathbf{B}) = \iint_\Omega A(\mathbf{s})I_\nu(\mathbf{s})e\left[i2\pi\nu\left(\frac{1}{c}\mathbf{B}\cdot\mathbf{s} - \tau_i\right)\right] d\Omega\, d\nu$$

In one dimension, the vectors become scalar quantities. The most important such change is that $\mathbf{B}\cdot\mathbf{s}$ becomes $B \times s$. Also, $\sin\theta$ becomes θ, and the double integral becomes a single integral. The resulting equation is:

$$R(B) = \int A(\theta)\, I_\nu(\theta) \exp\left[i2\pi\nu_0\left(\frac{1}{c}B\cdot\theta\right)\right] d\theta$$

© Springer International Publishing AG, part of Springer Nature 2018
T. L. Wilson, S. Hüttemeister, *Tools of Radio Astronomy – Problems and Solutions*, Astronomy and Astrophysics Library,
https://doi.org/10.1007/978-3-319-90820-5_25

This can be further simplified by noting that $v_0 \frac{1}{c} B$ is $\frac{B}{\lambda}$, so that we obtain Eq. (9.46):

$$R(u) = \int A(\theta) \, I_\nu(\theta) \exp\left[i \, 2\pi \left(\frac{B}{\lambda} \right) \cdot \theta \right] d\theta$$

Here, $D = B$ is the baseline in number of wavelengths, $\frac{B}{\lambda} = u$ and R is the response.

In the following, we set $A(\theta) = 1$. From the definitions of Fourier Transform (FT) pairs, Eq. (9.46) is the expression of an FT in the image plane, whereas

$$I(\theta) = \int R(u) \exp\left[-i \, 2\pi \left(\frac{B}{\lambda} \right) \cdot \theta \right] du = \int R(u) e\left[-i \, 2\pi \, (u) \cdot \theta \right] du$$

is the inverse transform, converting data in the u plane to an image. Thus, u and θ are the variables that describe the quantities in the two Fourier planes.

In Fig. 9.4, In panel (a), on the top left, the phase and the amplitude are plotted separately. So the expression for a 'point source' (i.e. a source whose angular size is smaller than the highest resolution of the instrument), shown on the right, which is offset from the origin is:

$$R(B) = A(\theta) \, I_\nu(\theta) \exp\left[i \, 2\pi \left(\frac{B}{\lambda} \cdot \theta_0 \right) \right]$$

Thus there is a phase offset, but a constant amplitude. The phase offset is $\frac{B}{\lambda} \times \theta_0 = u \times \theta_0$. In the u plane on the left, the offset is the inverse of θ_0 in radians. With increasing u, the phase will increase. The product of $u \cdot \theta_0$ will reach unity when u equals $\frac{1}{\theta_0}$. The value of θ_0 is given in radians; the conversion to arc seconds is given in the caption of figure 9.1.

In panel (b), on the right, the source is offset from the origin and extended. The phase offset is as in the topmost part of this figure. The source Full Width to Half Power (FWHP) size is d. The shape is a Gaussian. Then the FT, needed to produce the response, is another Gaussian of size proportional to $\frac{1}{4 \ln 2 \, d}$.

In panel (c), There are two point sources, with intensities R and unity. The response is

$$R(B) = R \exp\left[i \, 2\pi \left(\frac{B}{\lambda} \cdot \theta_0 \right) \right] + \exp\left[i \, 2\pi \left(\frac{B}{\lambda} \cdot \theta_1 \right) \right]$$

where $\theta_0 - \theta_1 = S$ in the x direction. In the u plane, if $u = 0$, the response is $1 + R$. At a u value of k_3/S, which corresponds to an angular distance of $S = \frac{1}{2}''$, the terms subtract. The normalized response reaches a minimum at this location. The phase offset reaches the value of $\frac{1}{2}$ when the angular offset is $\frac{1}{1+R}$ since the response is proportional to the intensities.

In panel (d), the situation is the same as in (c), but with two equally extended sources. Thus the analysis is a combination of (b) and (c).

3. Use coordinates x (representing θ) on the left, and u. Then
$I(x) = \int V(u)e^{-2\pi iux}du$ for all parts of this problem.
(a) $I(x) = \int_0^{u_0} V(u)e^{-2\pi iux}du$
$I(x) = 1/(2\pi ix)\left(1 - e^{-2\pi iu_0x}\right) = 2u_0e^{-2\pi iu_0x}(\sin 2\pi u_0x/2\pi u_0x)$
For large u_0, the zero crossing is at small x values. These negative sidelobes will distort the image in the x coordinate.
Panel (b): $I(x) = \int_{u_{min}}^{u_{max}} V(u)e^{-2\pi iux}du =$
$\int_0^{u_{max}} V(u)e^{-2\pi iux}du - \int_0^{u_{min}} V(u)e^{-2\pi iux}du$
$\int_0^{u_{max}} V(u)e^{-2\pi iux}du - \int_0^{u_{min}} V(u)e^{-2\pi iux}du =$
$\int_0^{u_{max}} V(u)e^{-2\pi iux}du - \left[2u_0e^{-2\pi iu_0x}(\sin(2\pi u_0x)/2\pi u_0x)\right]$
The second term gives rise to a 'Negative Bowl' in the x coordinate, which is the image plane.
Panel (c): This is a convolution of the 'picket fence' found in problem 6 of chapter 4 of this volume, and a function which has a smaller correlated intensity at larger spacings. From Table A.3 in Appendix A of 'Tools', the Fourier Transform of the Picket Fence, $III(x)$ is another Picket Fence, in which the spacings in the x coordinate are inversely proportional to the spacings in the u coordinate. As shown in problem 6 of chapter 4, the samples in the u coordinate must be fine enough so that the 'Grating Response' in the x coordinate is clearly separated from the image.

4. (a) From Eq. (6.78), $\theta = 1.02\lambda/D = 1.02 \times 3\,\text{cm}/100\,\text{cm} = 3.06 \times 10^{-2}\text{rad} = 1.76°$. Thus the antenna beam is much larger than the diameter of the Sun.
(b) Use the relation in Problem 1 of this chapter:
$T_A = \eta_B T_0 \frac{\theta_s}{\theta_s+\theta_A} = (0.7) \times (5800)\left(\frac{30^2}{30^2+105.6^2}\right) = 303\,\text{K},$
$T_B = T_A/\eta_B = 433\,\text{K}.$

(c) The fringe spacing is λ/B, where B is the baseline projection in the direction of the source. At transit, this is $3\,\text{cm}/10{,}000\,\text{cm} = 3 \times 10^{-4}\,\text{rad} = 61.9''$.
$V(u) = \int_{\theta_s/2}^{-\theta_s/2} A I_0 e^{2\pi iux}dx$. The antenna response A can be taken as a constant, A_0. Thus
$V(u) = A_0 I_0 \frac{\sin \pi u \theta_s}{\pi u}$ with $u = B/\lambda$, where B is the baseline projection in the direction of the source.
For transit, we get $u = 100 \times 10^2\,\text{cm}/3\,\text{cm} = 3.33 \times 10^3\,\text{rad}$. In addition, the source size in radians is $\theta_s = 8.73 \times 10^{-3}$. For these values, we get
$V(u) = A_0 I_0 \frac{\sin 91.33}{10460} = -A_0 I_0 \frac{0.221}{10460}.$
Thus the source is heavily resolved out by this spacing.
(e) For a 2 m separation, $B\theta_s = 66.67$.

5. Parts **(a)** and **(b)** follow the analysis used in problem 4. For part **(c)**, we have
$V(u) = A I_0 \left(\int_{\theta_1+\theta_2/2}^{\theta_1-\theta_2/2} e^{2\pi iux}dx + \int_{-\theta_1+\theta_2/2}^{-\theta_1-\theta_2/2} e^{2\pi iux}dx\right)$, or
$V(u) = 2 A_0 I_0\theta_2 \left(\frac{\sin \pi u \theta_2}{\pi u\theta_2}\right) \cos(2\pi u \theta_1).$
With u is as in the previous problem, $\theta_2 = 50''$, $\theta_1 = 45''$.

For these values, $\left(\frac{\sin \pi u \theta_2}{\pi u \theta_2}\right) = 0.228$. The regions are somewhat resolved.
While $\cos 2\pi u \theta_1 = -0.158$. As the source is tracked, the response will vary with the projected baseline. This source is barely resolved into two regions. At the shortest spacing, the source is unresolved.

6. Parts **(a)** and **(b)** follow the analysis in Problem 4 and 5. For part **(c)**, we have
$V(u) = A I_0 \left(\int_0^\infty e^{-4\ln 2\, x^2/\theta^2} e^{2\pi i u x} dx\right)$.
Use Tables A.3 (first entry) in 'Tools' to obtain the result
$V(u) = A I_0 e^{-\theta^2 u^2/4\ln 2}$.
For $\theta = 2.5' = 7.24 \times 10^{-4}$ rad, at transit, the visibility is 0.121 of the maximum value. So this source is rather resolved.
For an antenna spacing of 2 m, the visibility is determined by the value $B\theta_s = 66.67$, so
$V(u) = A I_0 e^{-(7.24\times10^{-4})^2 \times (66.67)^2/4\ln 2} = 0.9992$. Thus, the source is unresolved.
When considered in one dimension, the analysis is very similar to that used for Cygnus A, although the two-dimensional structures are quite different. Parts **(a)** and **(b)** are the same. For part **(c)**, we have
$V(u) = 2 A_0 I_0\theta_2 \left(\frac{\sin \pi u \theta_2}{\pi u \theta_2}\right) \cos 2\pi u \theta_1$.
$u = 3333.3$, as in the previous problem. Now for Cas A, $\theta_2 = 1' = 2.9 \times 10^{-4}$ rad, and $\theta_1 = 2.75' = 7.97 \times 10^{-4}$ rad.
For these values, $\left(\frac{\sin \pi u \theta_2}{\pi u \theta_2}\right) = 0.035$. The regions are rather resolved.
The term $\cos 2\pi u \theta_1$ has a value 0.944.
For the 2 m spacing, the source is unresolved.

7. We make use of Eq. (9.6) and note that all of the terms, especially I and A are taken to be frequency independent. Then we have

$$R(\mathbf{B}) = \int\int A(\mathbf{s})I_\nu(\mathbf{s})e^{2\pi i\nu}\left[(\mathbf{B} \times \mathbf{s} - \tau_i)\right] d\mathbf{s} d\nu .$$

We next integrate the frequency term from $\nu_0 + \Delta\nu_{IF}$ to $\nu_0 - \Delta\nu_{IF}$ and set $\Delta\tau = \tau_g - \tau_i$

$$R(\mathbf{B}) = A I_0 \Delta\nu_{IF} \frac{\sin(\pi \Delta\nu_{IF}\Delta\tau)}{(\pi \Delta\nu_{IF}\Delta\tau)} e^{2\pi i\nu_0\Delta\tau} \int \frac{\sin(\pi \mathbf{B} \times \mathbf{s} - \tau_i)}{\pi \mathbf{B} \times \mathbf{s} - \tau_i} d\mathbf{s} . \qquad (25.1)$$

Ignoring the IF frequency term, the value of $R(\mathbf{B})$ will be largest if the term in brackets in the integral is unity. This is the case for
$2\pi i\nu (\mathbf{B} \times \mathbf{s}) - \tau_i = 0$.
By adjusting τ_i, this can be so for the upper sideband, which has a phase of $\phi_1 = 2\pi(\nu_{LO} + \nu_{IF})\tau_g$
but *not* for the lower sideband, which has a phase of $\phi_1 = 2\pi(\nu_{LO} - \nu_{IF})\tau_g$.
Thus the interferometer can provide a natural way to separate sidebands.

8. (a) The filters must be placed in each IF *before* the multiplication. After filtering, each separate frequency channel is multiplied. The filters must be phase matched to prevent decorrelation of the signal.

From the previous problem we want the term
$\sin(\pi \Delta \nu_{\text{IF}} \Delta \tau)/(\pi \Delta \nu_{\text{IF}} \Delta \tau)$ to be a maximum. This requires that $\Delta \nu_{\text{IF}} \Delta \tau \ll 1$.

(b) Arrange the phases so that $\Delta \tau = 0$, in the center channel. Then the first $N/2$ channels will have a negative delay, the last $N/2$ channels a positive delay. The spectra from the phase center are symmetric, while those off the phase center are asymmetric. The Fourier transform of this cross correlation gives $N/2$ spectral data points, and $N/2$ phases.

9. Adding the voltages, $(V_1 + V_2)^2 = V_1^2 + 2V_1 V_2 + V_2^2$.
Subtracting, $(V_1 - V_2)^2 = V_1^2 - 2V_1 V_2 + V_2^2$.
Subtracting these two, we have $4V_1 V_2$. One half of the time is spent on each phase, so the noise is $\sqrt{2}$ larger than a multiplying interferometer. An additional factor of $\sqrt{2}$ in noise arises from subtracting two noisy signals.

10. The one dimensional version of Eq. (9.46) is stated in problem 2. As a first order approximation, we assume that the factor A is independent of angle, and take this outside the integral, so:

$$R(B) = A \int I_\nu(\theta) \exp\left[i 2\pi \nu_0 \left(\frac{1}{c} B \cdot \theta\right) \right] d\theta$$

The limits on the integral are θ_0 and 0. The result is

$$R(B) = A I_0 \left(e^{2\pi i (B/\lambda) \theta_0)} - 1\right)$$

Extracting factors $e^{\pi i (B/\lambda) \theta_0}$, and $\frac{B}{\lambda} = 1/\left(\theta_{\text{geom}}\right)$, we have

$$R(B) = A I_0 \theta_0 e^{i \pi \theta_0/\theta_{\text{geom}}} \times \left(\sin \pi \theta_0/\theta_{\text{geom}}\right)/\left(\pi \theta_0/\theta_{\text{geom}}\right)$$

The one dimensional version of Eq. (9.46) does not include a frequency term. Including this, and assuming that the source has no frequency dependence, we take I outside the integral. Then we have:

$$\int R(B) \, d\nu = A I_\nu(\theta) \int_{\nu_l}^{\nu_h} e^{\left[i 2\pi \nu_0 \left(\frac{1}{c} B \cdot \theta\right) \right]} d\theta$$

This integral depends only on the exponential term, so deal with this and ignore the rest of the equation. Evaluating the upper and lower limits, we have: $(1)/(\theta_0 B c) \times \left(e^{2\pi i (\theta_0 B c \nu_h)} - e^{2\pi i (\theta_0 B c \nu_l)}\right)$ The term involving ν in the exponential determines the phase offset for the position offset from the center of phase, $\Delta \phi$. This is $e^{2\pi i (\theta_0 B (c/\nu_h))} - e^{2\pi i (\theta_0 B (c/\nu_l))}$ Setting $1/\nu_h = 1/(\nu_l + \Delta \nu) = (1/\nu_l) \times (1 - \Delta \nu/\nu_l$ to first order, we have a difference in phase in these two terms of

$2\pi(\theta_0 B\ (c/\nu_l\ \Delta\nu/\nu_l)$ Using $B/\nu_l = 1/\theta_{\text{geom}}$, we have: $2\pi\ (\theta_0/\theta_{\text{geom}})\ (\Delta\nu/\nu_l)$
Since by assumption $\nu_l \approx \nu_h$, so can set $\nu_l = \nu$.
From this, on axis there is no phase difference as a function of wavelength. This is
the 'white fringe' in optics.

11. Use equation (9.27), with $M = 1$:
$$\Delta S = \frac{2kT_s}{A_e\ \lambda^2\ \sqrt{n(n-1)\Delta\nu\tau}}.$$
With $n(n-1) \approx n^2$, $\Delta\nu = (\nu/c)\Delta V$, and $S = 2.65\frac{\Delta T_B\theta^2}{\lambda^2}$, where λ is in cm and θ
is in arcmin, we have $\Delta S = \frac{8k}{\pi}\frac{T_{\text{sys}}\lambda^{0.5}}{n\,D^2\,\sqrt{\Delta V\tau}}.$
In addition, $\theta = \lambda/B_{\text{max}}$, so we have

$$\Delta T_B = \frac{8k}{2.65\,\pi}\frac{\lambda^{0.5}\,T_{\text{sys}}\,B_{\text{max}}^2}{n\,D^2\,\sqrt{\Delta V\tau}}$$

12*. (a) For the single dish, $A_{\text{sd}} = \frac{\pi}{4}D^2$, while each interferometer dish has $A_i = \frac{\pi}{4}d^2$, and the number of telescopes is $N = (D/d)^2$.
(b) For a single pointing, the image will have an angular size of λ/d. For a single
dish the beamsize is λ/D, so need to sample $N = (2D/d)^2$ beams. Per position the
integration time is the total time divided by N, or the integration time available for
each position is reduced by $(d/2D)^2$. This is not so for the interferometer since the
entire time is spent on all of the positions. The advantage for the interferometer is
due to the larger number of receivers.
(c) The angular resolution of the individual interferometer antennas is $\theta = \lambda/d$.
Take the size of the extended region to be $\Theta \gg \theta$. To completely sample this region,
one must have $N = (2\Theta/\theta)^2$ pointings. If there is a total time T allotted to image
the source, the time spent on each pointing is $T(\lambda/(\Theta d)^2$. From the last problem,
the temperature sensitivity is $\Delta T \sim 1/d^2\sqrt{t}$. Using the expression for total time,
the noise becomes $\Delta T \sim 1/(d\sqrt{T})$, where T is the total integration time. Thus
the RMS noise is proportional to a factor $1/d$ rather than the factor $1/d^2$.

13. Use Eq. (9.29) in 'Tools', which is:

$$\Delta T_B = 838.0\frac{M\,\lambda^2\,T'_{\text{sys}}}{A_e\theta_B^2\,\sqrt{N\,t\,\Delta\nu}}$$

set $M = 1$. The proposed measurement will use a $0.15\,\text{km s}^{-1} = 57.68\,\text{kHz}$
resolution. The value of A_e is $113.6\,\text{m}^2$. Assuming 0.7 for the antenna efficiency, η_e,
the value of $A_e = 79.2\,\text{m}^2$, N=49 × 50 /2, so that the time needed is 0.7 s. For a $10''$
beam, the antennas would have to have a maximum separation of only 50 meters.
So only 16 ALMA 12-m antennas could be fitted into a 50 m circular region. So this
is not practical. For a $1''$ resolution, the integration time is 100 times larger. There
are web sites for ALMA sensitivity which allow a check of this value.

14. $\theta = \lambda/D = 6.0\,\text{cm}/227 \times 10^5\,\text{cm} = 2.64 \times 10^{-7}\text{rad} = 0.0545''$.

$S(\text{Jy}) = 2.65\frac{T\theta^2}{\lambda^2}$, where θ is in arcmin and λ is in cm. Converting to units of arc sec and flux density in mJy, we have $1.36S[\text{mJy}] = T\frac{\theta('')^2}{\lambda^2}$. Applying this relation, we find that the uncertainties are equivalent to a main beam brightness temperature of $T = 16.4\,\text{K}$. Thus, the emission could arise from a thermal source.

15. Use Eq. (9.29) as given in problem 13. For the JVLA, the values are $\Delta V = 0.1\,\text{km s}^{-1}$, $\Delta v = 7.67\,\text{kHz}$, receiver noise is $100\,\text{K}$ (including the emission from the earth's atmosphere), the area of each dish is $490\,\text{m}^2$, and the efficient is assume to be 0.5. So for a 1 s integration:

$$\Delta T_B = 838 \times \frac{13^2 \times 100}{240 \times 3^2 \times 52.32} \times \frac{1}{\sqrt{t}} = 126\frac{1}{\sqrt{t}}$$

The actual noise temperature in the D array is $56\,\text{K}$, since the beam size is $4.5''$, not $3''$.

For the 50 antenna ALMA at 115 GHz, for $\Delta V = 0.1\,\text{km s}^{-1} = \Delta v = 38.4\,\text{kHz}$, the receiver noise is about $140\,\text{K}$ (including the emission from the earth's atmosphere), the area of each dish is $113\,\text{m}^2$, and the dish efficiency is 0.7. So for a 1 s integration:

$$\Delta T_B = 838 \times \frac{2.6^2 \times 140}{113 \times 0.7 \times 3^2 \times 217} \times \frac{1}{\sqrt{t}} = 5.2\frac{1}{\sqrt{t}}$$

Thus, ALMA is more sensitive, due the higher antenna efficiency, shorter wavelength and larger bandwidth in kHz for the same velocity resolution. In the context of the astrophysics, however, the JVLA has advantages for the measurement of most synchrotron sources, since these are more intense at lower frequencies. ALMA has advantages for the measurement of dust emission and molecular lines, since these more intense at higher frequencies. Such additional factors must be taken into account.

The ALMA sensitivity calculator gives $5.6\,\text{K}$ for this configuration. This calculator is to be found at:
https://almascience.eso.org/proposing/sensitivity-calculator.
The JVLA calculator is to be found at: https://obs.vla.nrao.edu/ect/.

Chapter 26
Solutions for Chapter 10: Emission Mechanisms of Continuous Radiation

1. (a) The angular diameter is $\theta = 0.1/(1.46 \times 10^6) = 6.84 \times 10^{-8}$ rad $= 2.4 \times 10^{-4}$ arcmin $= 1.4 \times 10^{-2}$ arcsec. The corresponding solid angle is $\Omega = 3.67 \times 10^{-15}$ steradians. The flux is given by $S = 2kT_B\Omega/\lambda^2 = 6.0 \times 10^{-4}$ Jy. If a Gaussian brightness distribution is assumed instead (doubtful for an asteroid), the result is $S = 2.65\,T_B\theta[']^2/\lambda[\text{cm}]^2 = 8.65 \times 10^{-4}$ Jy.
(b) For a Gaussian-shaped source, we have $T_{MB} = \theta_{source}^2/(\theta_{source}^2 + \theta_{beam}^2)\,T_B$. Here, $\theta_{source} \ll \theta_{beam}$, so $T_{MB} = 100\,\text{K}\,\left(\frac{1.4 \times 10^{-2}}{12}\right)^2 = 1.4 \times 10^{-4}$ K.

(c) We use a total power expression for the noise, combining $\text{NEP} = 2kT_{noise}\sqrt{\Delta\nu}$ (see problem 17 of chapter 8) with $T_{RMS} = T_{noise}/\sqrt{t_{int}\,\Delta\nu}$ to get $T_{RMS} = \text{NEP}/(2k\,\Delta\nu\,\sqrt{t_{int}}) = 2 \times 10^{-15}/(2 \times 1.38 \times 10^{-23} \times 20 \times 10^9\,\sqrt{t_{int}})$ This y gives a result $T_{RMS} = 3.6 \times 10^{-3}/\sqrt{t_{int}}$ K. The factor of 2 is explained by the need for beam switching to remove the emission from the earth's atmosphere. Also, the source intensity must be halved because $\eta_{MB} = 0.5$. To get a S/N ratio of 5:1, the integration time is $\sqrt{t_{int}} = (3.6 \times 10^{-3})/(1.4 \times 10^{-5}) \rightarrow t_{int} = 6.6 \times 10^4$ s $= 18.4$ h.

2. (a) If the line-of-sight depth equals the diameter, it is $0.0242\,\text{pc} = 7.5 \times 10^{16}$ cm.
(b) The column density is $N(\text{H}_2) = (7.5 \times 10^{16}\,\text{cm})(10^7\,\text{cm}^{-3}) = 7.5 \times 10^{23}\,\text{cm}^{-2}$.
(c) From the relation in the statement of this problem, we have $S_\nu = 8.2$ Jy.

(d) Application of $T_{MB} = 0.377\,S_\nu\,\lambda^2/\theta_0^2$ with $\theta_0 = \sqrt{\theta_{source}^2 + \theta_{beam}^2} = 14.1''$ gives $T_{MB} = 0.95$ K, which is indeed $\ll T_{dust}$.
(e) We take $\theta_0 = \theta_{source} = 10''$. Thus, $T = 1.88$ K and $\tau = 1.88/160 = 0.012$.
(f) Calling $x[\mu\text{m}]$ the wavelength where $\tau = 1$, we have $\tau(1.3\,\text{mm})/\tau(x) = 0.012 = x^4/(1300)^4$. So, $x = 430\,\mu\text{m}$.

3. (a) From Fig 10.1 $\nu_0 \approx 0.9$ GHz.
(b) Using the given formula, we have $0.9 = 0.3045 \times (8300)^{-0.643}\,(\text{EM})^{0.476}$, so $\text{EM} = 1.92 \times 10^6\,\text{cm}^{-6}\,\text{pc}$.

© Springer International Publishing AG, part of Springer Nature 2018
T. L. Wilson, S. Hüttemeister, *Tools of Radio Astronomy – Problems and Solutions*, Astronomy and Astrophysics Library,
https://doi.org/10.1007/978-3-319-90820-5_26

Table 26.1 Optical depths and temperatures for free–free emission from Orion A (problem 5)

Frequency ν (GHz)	Optical depth τ_{ff}	$T_B(K)$	$T_{MB}(K)$
23	3.1×10^{-3}	26	16
90	1.8×10^{-4}	1.5	1.4
150	6.0×10^{-5}	0.5	0.5
230	2.5×10^{-5}	0.2	0.2

(c) The diameter for a (Gaussian) source is $D = 0.364 \, \text{pc} = 1.12 \times 10^{18} \, \text{cm}$. given by $\langle n_{RMS} \rangle = \sqrt{EM/D} = (1.92 \times 10^6 \, \text{cm}^{-6} \, \text{pc})/(0.364)^{0.5} = 2.3 \times 10^3 \, \text{cm}^{-3}$. Larger values of density are obtained from an analysis of optical measurements. This is a sign for 'source clumping', this indicates that the source has small scale structure.

4. (a) The actual source temperature is determined by $T_B = ((\theta_s^2 + \theta_B^2)/\theta_s^2) \times T_{MB} = 1.08 \, T_{MB} = 25.9 \, \text{K}$.
(b) We have $25.9 \, \text{K} = 8300 \, \text{K} \, \tau_{ff}$, so that $\tau_{ff} = 3.1 \times 10^{-3}$.
(c) Solving $\tau_{ff} = 3.1 \times 10^{-3} = 0.08235 \times (8300)^{-1.35} \times (23)^{-2.1}$ EM, we find EM $= 5.3 \times 10^6 \, \text{cm}^{-6} \, \text{pc}$, and (using the formula from the last problem) $\nu_0 = 1.46 \, \text{GHz}$. The value for EM is about 2.8 times larger than the value obtained in Problem 3. The present value is obtained without assumptions about source structure, which is an advantage. However, at high frequencies (i.e. far beyond the turnover frequency ν_0), the emission becomes rather weak and noise can cause large errors.

5. (a) Use EM $= 5.3 \times 10^6 \, \text{cm}^{-6} \, \text{pc}$ in Eq. (10.37) in 'Tools' (see Problem 4(**c**)), so $\tau_{ff} = 2.243 \, \nu^{-2.1}$. The values are collected in Table 26.1.
(b) The results are listed in Table 26.1.
(c) T_{MB} is obtained by $T_{MB} = (\theta_s^2/(\theta_s^2 + \theta_b^2)) \, T_e \tau_{ff}$. See Table 26.1 for the results. At 5 GHz, the 30 m telescope beam is $9'$, at 10 GHz, $4.5'$, at 23 GHz, $1.95'$. At higher frequencies, the source is much larger than the beam, so corrections for beam size are small.

6. (a) One can obtain a solution by setting the results for the dust parameters of Orion KL from Problem 2, namely, $\tau_{dust} = 7 \times 10^{-21} \, b \, N_H \, \lambda[\mu m]^{-2})$ equal to those of the last problem. The parameters include beam filling factors and the dust temperature and column density of the dense core as well as the optical depth, electron temperature and free–free optical depth of the HII region. Therefore, the result is specific to Orion A and Orion KL, as observed with a specific telescope. The plot in Fig. 10.1 shows that significant dust emission occurs at frequencies larger than $\sim$130 GHz. The simplest approach is to use numerical values between 130 GHz and 230 GHz. For Orion KL, these temperatures are equal at about 180 GHz (T_{MB}(Orion A) $= 0.34 \, \text{K}$).
(b) Only the beam filling factors differ. A typical 10-m or 12-m radio telescope has a beam size that is about three times larger than the IRAM 30-m beam. Since the Orion KL nebula has a far smaller angular size than the Orion HII region, the main beam temperature from dust emission will be much smaller than with

the 30 m telescope. *Thus, the frequency at which the temperatures are equal will be higher.* Since dust emission increases rapidly with frequency while free–free emission decreases about as rapidly, the increase in frequency is modest, however. Given the larger beam sizes of the 10-m or 12-m telescopes, the main factors are the source sizes, EM and dust parameters. The results are in Table 26.2. temperatures will be equal at $\sim$240 GHz, where they both reach about 0.17 K.

7. (a) The turn over frequency is Eq. (10.37) and is also given in the solution to problem 3(b) of this chapter. The Emission Measure, EM for W3(OH) is $EM = 1.8 \times 10^9 \, \text{cm}^{-6}$ pc, for a size of $2''$ at 1.88 kpc, the linear size in pc is 1.84×10^{-2}, and the RMS electron density is $3 \times 10^5 \, \text{cm}^{-3}$.
(b) The source sizes are the same, so we use the values of T_B, which is the temperature without taking the beam size into account. The dust and free-free emission temperatures, as a function of frequency are in the following table. The relevant equation for dust emission is Eq. (10.6), given in the statement of problem 2 of this chapter. As before, we set the value of Z=Z$_\odot$, N(H)=$10^{24} \, \text{cm}^{-2}$ and $\theta = 2''$. The conversion from flux density to T_B makes use of the result of problem 10 in chapter 1. The results are given in Table 26.2.
(c) Even though the beam size of the 30-m is smaller, the beam sizes of both telescopes are larger than the source sizes and their separation. Note we have given the brightness temperatures. For an estimate of the observed temperatures, the values given above must be multiplied by the factor $\frac{\theta_s^2}{\theta_s^2+\theta_B^2}$, where θ_s^2 is the source size and θ_B^2 is the beam size.

8. (a) The background photon energy is $E = h\nu = hc/\lambda = 1.24 \times 10^{-15}$ erg. The 5 keV electrons have $E = (1.6 \times 10^{-12} \, \text{erg/eV}) \times 5 \times 10^3 = 8 \times 10^{-9}$ erg. Thus the electrons have vastly more energy.
(b) With an energy density of $1.3 \times 10^{-12} \, \text{erg cm}^{-3}$, the number of photons is $\sim$ 400 cm^{-3}, while the number of electrons is about $10^{-2} \, \text{cm}^{-3}$. So only a few photons will be affected.
(c) The energy density of the hot electrons, before the interaction, is $\sim 8 \times 10^{-11} \, \text{erg cm}^{-3}$, or more than 60 times the energy density of the background radiation. If the result of the interaction is equipartition, the background radiation gains energy and most photons are promoted to higher energies. So, longer-wavelength photons are shifted to shorter wavelengths, resulting in a deficit of photons below the peak of background radiation at 1.6 mm (decrease in temperature), but in an excess shortward of 1.6 mm (increase in temperature).

Table 26.2 Temperatures for thermal dust and free-free emission from W3(OH) (problem 7)

Frequency ν (GHz)	$T_{B,dust}(K)$	$T_{B,\text{HII Region}}$ (K)
200	54	94
230	72	70
250	85	59

Table 26.3 The brightness temperatures of Cassiopeia A (problem 9)

Wavelength λ (m)	Flux density S (Jy)	$T_B(K)$
3	1.9×10^4	2.4×10^7
0.3	3.4×10^3	4.2×10^4
0.03	750	8.1×10^1
0.003	$\sim$130	1.6×10^{-1}

9. First, we must calculate the solid angle Ω: $\Omega = 2\pi \int_{4.5'}^{5.5'} \sin\theta \ d\theta = 2\pi \left(4.23 \times 10^{-7}\right) = 2.66 \times 10^{-6}$ steradians. Then, consistent units must be used in $S = 2k \, T_B \Omega / \lambda^2$. Flux densities are estimated from Fig. 10.1, given in the statement of this problem. The results are given in the table below. See Kantharia et al. 1998, ApJ **506**, 758 for references (Table 26.3).

10. The power emitted by a single electron is Eq. (10.66), in 'Tools'. In a simplified form this is: $P_e = C E^2 B^2$, where C is a constant, E the energy of the electron and B the magnetic field. With the given constant-energy distribution, we obtain the integrated power of the ensemble as $P = \int_{E_{min}}^{E_{max}} N_0 C E^2 B^2 dE = \frac{1}{3} N_0 C B^2 \left(E_{max}^3 - E_{min}^3\right)$.
In Eq. (10.91) of 'Tools', $\delta = 0$, so $n = -1/2$. Thus from Eq. (10.96) in 'Tools', the emissivity of this ensemble of electrons between energies E_{min} and E_{max} is $\varepsilon \sim \nu^{1/2}$, i.e. the spectral index is $1/2$. [Research has shown that the most well-known flat spectrum Synchrotron source, Sgr A*, has a flat spectrum because the emission is optically thick. For this source, the radiation becomes optically thin at frequencies higher than 345 GHz. Very Long Baseline measurements of this source must be carried out at frequencies above 230 GHz to probe the inner parts of this source.]

Chapter 27
Solutions for Chapter 11: Some Examples of Thermal and Nonthermal Radio Sources

1. (a) The emission measure is defined as $EM = \int n_e^2 dr$. Thus

$$EM = \int_{r_0}^{\infty} \left[1.55 \left(\frac{r}{r_0} \right)^{-6} + 2.99 \left(\frac{r}{r_0} \right)^{-16} \right]^2 10^{16} \, dr = 1.65 \times 10^8 \, \text{cm}^{-6} \, \text{pc}.$$

(b) Inserting the numbers, the optical depth is $\tau = 8.235 \times 10^{-2} \times (10^6)^{-1.35} \times (1.65 \times 10^8)\nu^{-2.1}$. The brightness temperature is given by $T_B = 10^6 (1 - e^{-\tau})$. Table 27.1 lists the results.

2. The average distance to the planet Jupiter is 5.2 Astronomical Units, or 7.8 $\times 10^{13}$ cm. The distance of 10 parsecs in cm is 3×10^{19} cm, so the ratio of distances, squared, is 6.8×10^{-12}. If the flux density from Jupiter is 10^5 Jy, we would receive 0.7 μJy from such a hot Jupiter.

3. (a) The maximum wavelength is $\lambda_{\text{max}} = (0.28978/2.73)\,\text{cm} = 0.106\,\text{cm}$. The corresponding frequency is $\nu = 282\,\text{GHz}$.
(b) For the $T - \nu$ relation, we have $\nu_{\text{max}} = 2.73 \cdot 58.789\,\text{GHz} = 160.5\,\text{GHz}$, or $\lambda = 0.19\,\text{cm}$. The difference is caused by the weighting of the Planck relation in terms of λ and ν. Substituting values for 160.5 GHz, B_ν at the maximum for the cosmic background radiation is $B_{\nu,\text{max}}(2.73K) = 6 \times 10^{-14}/(e^{2.82} - 1) = 3.8 \times 10^{-15}\,\text{erg cm}^{-2}\,\text{s}^{-1}\,\text{steradian}^{-1}$.
(c) The integrated intensity is $I = \int_0^\infty B_\nu \, d\nu = \sigma T^4 = 3.2 \times 10^{-3}\,\text{erg cm}^{-2}\,\text{s}^{-1}$, and the energy density follows to $u = 4\pi I/c = 1.3 \times 10^{-12}\,\text{erg cm}^{-3}$.
(d*) The number density of photons $< n >$ is given by $< n >= \frac{4\pi}{c}$ $< I(T)/(h\nu) >$. For a black body, I_ν is given by the Planck function, B_ν. Thus, inserting Eq. (1.13) from 'Tools' (also given in the statement of problem 14 of chapter 1) and integrating over ν results in $< n >= 8\pi/c^3 \int_0^\infty \frac{\nu^2}{e^{h\nu/kT}-1} d\nu$.

© Springer International Publishing AG, part of Springer Nature 2018
T. L. Wilson, S. Hüttemeister, *Tools of Radio Astronomy – Problems and Solutions*, Astronomy and Astrophysics Library,
https://doi.org/10.1007/978-3-319-90820-5_27

Table 27.1 Optical depth and brightness temperature of the solar atmosphere (problem 1)

Frequency (GHz)	Optical depth τ	Brightness temperature (K)
0.1	13.5	10^6
0.6	0.32	2.7×10^5
1.0	0.11	1.0×10^5
10	8.6×10^{-4}	8.6×10^2

Substituting $x = h\nu/kT$, this becomes

$$< n >= \frac{8\pi}{c^3} \left(\frac{kT}{h} \right)^3 \int_0^\infty \frac{x^2}{e^x - 1} \, dx = \frac{8\pi}{c^3} \cdot 2.404 \cdot \left(\frac{kT}{h} \right)^3 = 20.3 \, T^3$$

For the 2.73 K background, this number density is $< n >= 412 \, \text{cm}^{-3}$.

(e) To first order, the ratio of the expression for radiation in the Rayleigh-Jeans regime to that in the Planck regime is $1 + \frac{1}{2} \frac{h\nu}{kT}$ For 4.8 GHz, this term is 1.04, i.e. the error is 4%. For 115 GHz, it is 101%, and for 180 GHz, it is 158%. Thus, above $\sim$70 GHz (error 60%), the Rayleigh-Jeans relation cannot be used to characterize the 2.73 K background.

4. Use the dependence of T on redshift, z ($z = \lambda_0/\lambda_f) - 1$), where λ_0 is the rest wavelength of a spectral line. Then we have:
$T = 2.73(1 + z)$, so for $z = 2.28$, T = 8.9 K, for $z = 5$, 16.4 K, and for $z = 1000$, T = 2733 K.
The gas ionized by the Big Bang recombines at a value $z \approx 1000$.

5. (a) The maximum frequency, from the $T - \nu$ relation (Eq. (1.25)), is $\nu_{max} = 2.73 \times 58.789 \, \text{GHz} = 160.5 \, \text{GHz}$, or $\lambda = 0.19 \, \text{cm}$.
(b) As stated in problem 3, the difference is the weighting.
(c) For this part and parts (d) and (e), the results are the same as in problem 3.

6. (a) The brightness temperature is $T_B = 6500(1 - e^{-\tau})$. So, the value of τ at 5 GHz is 7.7×10^{-5}. Assuming that $\tau \sim \nu^{-2.1}$, we have $\tau(23 \, \text{GHz}) = 3.1 \times 10^{-6}$. Then the brightness temperature should be 2×10^{-2} K. This level of emission is well below the sensitivity of a moderate sensitivity map.
(b) With the same scaling, we find $((\nu(\tau = 1))/5)^{-2.1} = 1/(7.7 \times 10^{-5})$ and thus $\nu(\tau = 1) = 55 \, \text{MHz}$.

7. Use the Rayleigh-Jeans relation to obtain the intensity, I_ν in $\text{W m}^{-2} \, \text{Hz}^{-1}$. This is $1.75 \times 10^{-14} \, \text{W m}^{-2} \, \text{Hz}^{-1}$. The solid angle of the surface of the star is $\Omega = \left(\frac{\pi \, 7 \times 10^{10}}{D} \right)^2$. Setting this product $i_n u \times \Omega$ equal to $10^{-32} \, \text{W m}^{-2} \, \text{Hz}^{-1}$, we have a value of D=$1.6 \times 10^{20}$ cm or 55 pc.

8. (a) Inserting the numbers, we have $S_\nu = 8.3 \left(\frac{(10^{10})(3.6 \times 10^{12})^2}{10^{36}} \right) \times (10)^{0.6} \times (1.6)^{0.1} \cdot (3)^{-2} = 0.50 \, \text{Jy}$.

(b) A flux density of 0.5 Jy corresponds to $T_A = 0.65$ K with the 100 m telescope at 10 GHz. The RMS noise, ignoring atmospheric contributions, is $\Delta T_{RMS} = \frac{2T_s}{\sqrt{B\, t_{int}}} = (4.47 \cdot 10^{-3})/\sqrt{t_{int}}$. For a $5\Delta T_{RMS}$, detection, one needs less far less than 1 s (nominally, ~ 0.2 s). Thus, the star is easily visible, however, confusion may be a significant effect.

(c) Using the given formula, we find $S_\nu = 2.7 \cdot 10^{-2} = 8.2 \left(\frac{n\, r_0^2}{10^{36}} \right) \times 23^{0.6} \cdot 2.0^{0.1} \times 7^{-2}$. Thus, the result is $n_0\, r_0^2 = 5.9 \times 10^{34}$ cm^{-1}.

(d) Given the electron density, we find that $r_0 = 2.4 \times 10^{12}$ cm or 0.16 AU. This could be a supergiant star.

9*. Evaluate the expression for $n_e(r) \cdot r^2$ in CGS units. This expression is $(\frac{\dot{M}}{4\pi\, v_w\, m_H\, \mu})$, where μ is the molecular weight in atomic units. Inserting the numerical values:

$$\frac{10^{-5} \times (2\, 10^{33}\mathrm{g})}{(3.15 \times 10^7\,\mathrm{s}) \times (1.257 \times 10^9\,\mathrm{cms}^{-1}) \times (1.67 \times 10^{-24}\,\mu\,\mathrm{g})} = 3.022 \times 10^{35}/\mu$$

Substituting this value into Eq. (10.7), we have:

$$S_\nu = 8.2 \left(\frac{3.022 \times 10^{35}}{10^{36}} \right)^{4/3} \cdot \left(\frac{\dot{M}[\mathrm{M}_\odot\,\mathrm{yr}^{-1}]}{v_{w[1000\,\mathrm{km s}^{-1})}\,\mu} \right)^{4/3} \times (v)^{0.6} \times (T_e)^{0.1} \times (D)^{-2}$$

The final result is

$$S_\nu = 1.7 \left(\frac{\dot{M}[10^{-5}\,\mathrm{M}_\odot\,\mathrm{yr}^{-1}]}{v_w[1000\,\mathrm{km\, s}^{-1}]\,\mu} \right)^{4/3} \times (v)^{0.6} \times (T_e)^{0.1} \times (D)^{-2}$$

Assuming that the outflow consists of hydrogen only, we set $\mu = 1$. Then, we find for the mass loss rate of the supergiant in problem 8(a):

$$\dot{M} = 6.3 \cdot 10^{-7}\, M_\odot\,\mathrm{yr}^{-1}$$

10. Taking $T = T_e$ and $v_w = 100\,\mathrm{km\, s}^{-1}$, we find $S_\nu = 1.70 \left(\frac{0.1}{0.11.0} \right)^{4/3} \cdot 10^{0.6} \cdot 3.1^{0.1} \cdot 7^{-2} = 0.15$ Jy. This is a source that should be detectable with the 100 m telescope and even more easily with the JVLA.

11. We must use the Rayleigh-Jeans relation of flux density and temperature, as given in the statement of problem 10 of chapter 1 (or Eq. (8.20) in 'Tools') to obtain the brightness temperatures of the Crab Nebula from the flux densities. Both estimated S_ν and T_b are given in Table 27.2. The Crab nebula is a non-thermal source, a synchrotron emitter. From Table 27.2, this source varies with by a factor of 4 over a frequency range of 100, so has a rather flat spectrum.

Table 27.2 Flux densities and brightness temperatures of the Crab Nebula (problem 11)

Frequency (GHz)	Flux density (Jy)	Brightness temperature (K)
0.1	1.5×10^3	2.0×10^6
10	4×10^2	54

12. A diameter of $5.5'$ at $D = 3$ kpc corresponds to 4.84 pc $= 1.49 \times 10^{19}$ cm. Use 1/2 of this for the distance travelled. The time elapsed is 332 yr $\approx 10^{10}$ s. Assume $R = vt$. Thus the speed is 7×10^8 cm s^{-1}, or 2.5% of the speed of light. The point source moves at $0.5''$ per year, using the assumptions above. This should be easy to measure with the JVLA, where motion should become visible after little more than a month.

13. (a) The flux density of Cas A at 100 MHz today is $\sim 2 \times 10^4$ Jy. Solving the equation given, we have $S_\nu(t) = S_\nu(t_0) (t/t_0)^{-4\delta/5}$. Setting $t_0 = 332$ yr, and $t = 100$ yr, we have $S_{100\,\text{MHz}}(100\,\text{yr}) = 2 \times 10^4 \times (0.301)^{-2.032} = 2.29 \times 10^5$ Jy, which is more than 11 times the present value. At that time, Cas A had a size of $D = 2 \times 7 \times 10^8$ cm s$^{-1} \times 100 \times 3.15 \times 10^7$ s $= 4.4 \times 10^{18}$ cm. This is equivalent to an angular size of $1.64'$ at 3 kpc.
(b) Use the Rayleigh–Jeans relation in problem 10 of chapter 1, with $S = 2.29 \times 10^5$ Jy and $\theta = 1.64'$. Then at 100 MHz, we obtain $T = 2.9 \times 10^9$ K.

14. $EM = (0.03\,\text{cm}^{-3})^2 \cdot (3000\,\text{pc}) = 2.7\,\text{cm}^{-6}$ pc. For this EM, τ is
$\tau = 0.08235\,(T_e)^{-1.35}(0.01)^{-2.1}(2.7) = 5.6 \times 10^4/(T_e)^{1.35}$
If T_e were 10^3 K or less, the absorption is large. If the free electrons have $T_e = 10^4$ K, $\tau = 0.22$, which is too small. There is a problem keeping a 10^3 K gas ionized, so it would seem that the electron density must be larger than average.
$EM = (0.03\,\text{cm}^{-3})^2 \cdot (3000\,\text{pc}) = 2.7\,\text{cm}^{-6}$ pc. For this EM, τ is
$\tau = 0.08235\,(T_e)^{-1.35}(0.01)^{-2.1}(2.7) = 5.6 \times 10^4/(T_e)^{1.35}$
If T_e were 10^3 K or less, the absorption will be larger. If the free electrons have $T_e = 10^4$ K, $\tau = 0.22$, which is too small. It seems unlikely that the $T_e = 10^3$ K for an ionized gas, so it is likely that the electron density must be larger than the average.

15. (a) We must differentiate W_{tot} (given in Eq. (10.118) in 'Tools', and given in the statement of this problem), then set B=B_{eq}, and set the result to zero:

$$dW_{\text{tot}}/dB_{\text{eq}} = 0 = -\frac{3}{2}\frac{G}{H}R^2 S_\nu \nu^n B_{\text{eq}}^{-5/2} + \frac{V}{4\pi}B_{\text{eq}}$$

Solving for B_{eq}, we obtain the given expression.
(b) The major point is the evaluation of the ratio $\frac{G}{H}$. This is

$$\frac{G}{H} = \left(\frac{\eta}{1-2n}\right)\left[(e/m^3 c^5)^{-0.5}(\nu_{\max}^{0.5-n} - \nu_{\min}^{0.5-n})\right] / \left[b\,(e^3/mc^2)\,(3/4\pi)^n\right] = \frac{A}{B}$$

where

$$A = \left(\frac{10}{-0.5}\right)\left[(4.8 \times 10^{-10})/(9.1 \times 10^{-28})^3 (3 \times 10^{10})^5)^{-0.5}\right.$$

$$\left. \times ((50 \times 10^9)^{-0.25} - (10^7)^{-0.25})\right)$$

simplifying, $A = (-20) \times \left[(2.62 \times 10^{19})^{-0.5}(-0.01567)\right]$ and for the other term, $B = \left[0.08\,(4.8 \times 10^{-10})^3/((9.1 \times 10^{-28})(3 \times 10^{10})^2))\,(3/4\pi)^{0.75}\right]$ This becomes: $B = \left[0.08\,(1.35 \times 10^{-22})\,(3/4\pi)^{0.75}\right]$ so the final result is: $\frac{A}{B} = 1.65 \times 10^{13}$

Converting S_ν from CGS units to Jy, we use $1\,\mathrm{Jy} = 10^{-23}$ in CGS units. For R we have $R(\mathrm{Mpc}) = 3 \times 10^{24} R(\mathrm{kpc})$, and for V we have $V(\mathrm{cm}) = (3 \times 10^{21})^3 V(\mathrm{kpc})$. With these values, have:

$$B_{eq} = \left(\left[6\pi\,(1.65 \times 10^{13})(3 \times 10^{24})^2\,R^2\,S_\nu(\mathrm{Jy})\,\nu(\mathrm{GHz})^{0.75}\right]\right.$$

$$\left. / \left[(3 \times 10^{21})^3\,V(\mathrm{kpc})\right]\right)^{0.286}$$

This gives: $B_{eq} = 1.2 \times 10^{-5}\,\left(\left[R^2\,S_\nu(\mathrm{Jy})\,\nu(\mathrm{GHz})^{0.75}\right] / \left[V(\mathrm{kpc})\right]\right)^{0.286}$

16. Using the equations given in problem 15, we have

$$B_{eq} = 1.2 \times 10^{-5} \times \left(\left[3.4^2 \times (2.1) \times (8.7)^{0.75}\right] / \left[(4.19 \times 10^3)\right]\right)^{0.286}$$

or $B_{eq} = 1.2\,10^{-5}\,(2.93 \times 10^{-2})^{0.286}$ The result is $B_{eq} = 1.2 \times 10^{-5} \times (0.364) = 4.3 \times 10^{-6}\mathrm{Gauss}$

The relativistic particle energy is $u_{\mathrm{rel.\,particle}} = 1.33\,u_{\mathrm{mag}} = (1.33) \times (\frac{1}{8\pi}) \times (4.3\,10^{-6})^2 = 9.8 \times 10^{-13}\mathrm{ergs\,cm}^{-3}$

17. The power (in Watts) is:

$$P = 4\pi \times (2.74 \times 10^{49}) \times (10^4 \times 10^{-26})/(0.01\,\mathrm{GHz})^{0.75} \int_{0.001}^{50} \nu^{-0.75}\,d\nu$$

This gives:

$$P = (1.1 \times 10^{27}) \times (4) \times \left((50)^{0.25} - (0.001)^{0.25}\right)$$

$$= 4.3 \times 10^{27}\,(2.66 - 0.18) = 1.1 \times 10^{28}\,\mathrm{W}$$

This gives: $P = 1.1 \times 10^{35}\,\mathrm{erg\,s}^{-1}$.

The total energy is much larger than the energy radiated per second. Thus synchrotron losses are small, the loss time is comparable to a Hubble time. The age of the lobes is $(7 \times 10^4 \text{pc})/(0.2c) = 10^6$ years.

18. (a). Using the simplest approach, namely, $v = c\,z\,H_0$, we obtain a radial velocity of $v = z \times c = (0.16) \times (3 \times 10^5 \text{ km s}^{-1}) = 4.8 \times 10^4 \text{ km s}^{-1}$ where H_0 is the Hubble constant of $70 \text{ km s}^{-1} \text{ Mpc}^{-1}$. From the Hubble constant given, get a distance of 480 Mpc.

Using the non-relativistic time variability, get that 1 light-month is a linear dimension of 0.28 pc. Then the source size in radians is

$\theta = (0.028\text{pc})/(4.8 \times 10^8 \text{ pc}) = 5.8 \times 10^{-11}$ rad, or 1.1×10^{-5} arcsec.

For the flux density given, the main beam brightness temperature at 1.5 cm in a $1'$ beam is 16.9 K. Using our result for problem 6, chapter 5, have

$$T_B = T_{MB} \left(\frac{60''}{1.2 \times 10^{-5 \,''}} \right)^2 = 4.3 \times 10^{14} \text{K}$$

This is much higher than the inverse Compton limit. This indicates that relativistic beaming effects must be important.

(b) Differentiating Eq. (11.61) in 'Tools', which is given in the statement of this problem and setting $V = c$, have

$$\cos(\theta) - V/c(\cos(\theta))^2 = V/c \sin(\theta)^2$$

or

$\cos(\theta) = v/c$. Thus as $v \to c$, the apparent speed approaches c/γ. For an apparent $v/c = 7$, have an angle of 16°. Thus, an apparent motion which is faster-than-light, but in which the actual velocity does *not* exceed the speed of light, is possible and likely.

Chapter 28
Solutions for Chapter 12: Spectral Line Fundamentals

1. (a) Use Eq. (12.4):

$$\frac{n_2}{n_1} == \frac{g_2}{g_1} \exp\left(-\frac{h\nu_0}{kT_e}\right)$$

with $g_2 = g_1$. Then

$$\frac{h\nu_0}{kT_e} = 0.00995$$

Thus the excitation temperature, T_e, is positive. For the first value, $n_1 = 1.01\,n_2$, $h\nu/T_e = 0.00995$. If $h\nu/k=0.068\,\mathrm{K}$ as in the case of the 21 cm line of atomic hydrogen, the value of $T_e = 6.8\,\mathrm{K}$. For a population ratio of $n_1 = 1.1\,n_2$, $h\nu/k = 0.0953$, so for the hygrogen line, $T_e = 0.7\,\mathrm{K}$. Given the presence of the 2.73 K microwave background, the value of T_e should not be less than this value, since atomic hydrogen is essentially a two level system (i.e. the next level above the ground state is at the energy equivalent of $9 \times 10^4\,\mathrm{K}$.

2. Using $n^* = A_{21}/\langle\sigma v\rangle$ results in critical densities as follows:
neutral hydrogen: $n^* = 2.85 \times 10^{-5}\,\mathrm{cm}^{-3}$; HCO^+: $n^* = 3 \times 10^5\,\mathrm{cm}^{-3}$ and CO: $n^* = 740\,\mathrm{cm}^{-3}$. The 21 cm transition is almost always thermalized ($n > n^*$), while high densities are required to thermalize the HCO^+ $1 \rightarrow 0$ line. CO is easier to thermalize than HCO^+, but still needs far higher densities than neutral hydrogen. Solving the given equation for n

$$n = \frac{A_{21}(T_b - T_{ex})}{T_0\langle\sigma v\rangle((T_{ex}/T_k) - 1)}.$$

With $\nu = 89.186\,\mathrm{GHz}$, $T_0 = 4.28\,\mathrm{K}$. Assuming that the radiation temperature is given by the cosmic background, $T_b = 2.73\,\mathrm{K}$, yields $n = 6.53 \times 10^4\,\mathrm{cm}^{-3}$.

© Springer International Publishing AG, part of Springer Nature 2018
T. L. Wilson, S. Hüttemeister, *Tools of Radio Astronomy – Problems and Solutions*, Astronomy and Astrophysics Library,
https://doi.org/10.1007/978-3-319-90820-5_28

For $\nu(CO) = 115.271\,\text{GHz}$, $T_0 = 5.53\,\text{K}$. With the other parameters the same as before, this gives $T_{\text{ex}} = 83\,\text{K}$, i.e. an excitation temperature that is already very close to the kinetic temperature. See problems 13 and 14 of chapter 14 for a discussion of negative excitation temperatures.

3.*(a) Inserting the given 'ansatz' for x in the differential equation yields $\alpha^2 + \omega_0^2 + \gamma\alpha = 0$. Solving for α gives $\alpha = (-\gamma/2) \pm \sqrt{(\gamma^2/4) - \omega_0^2}$. Assuming $\gamma^2/4 \ll \omega_0^2$, we have $\alpha = (-\gamma/2) \pm i\omega_0$. Thus: $x = x_0\, e^{-\gamma t/2}\, e^{-i\omega_0 t}$.

(b) The electric field $E(t)$ is given by: $E(t) = e^{-\gamma t/2}\, e^{-i\omega_0 t}$.

(c) According to the definition of the (inverse) Fourier transform (see e.g. 'Tools', Appendix A), we have

$$E(\nu) = \int_{-\infty}^{\infty} e^{-\gamma t/2}\, e^{-i\omega_0 t}\, e^{i2\pi\nu t}\, dt\ .$$

Using $\omega = 2\pi\nu$ and defining $u = (\omega - \omega_0)t$, we can solve this integral:

$$E(\nu) = 2\int_{0}^{\infty} e^{-\gamma t/2}\, e^{i(\omega - \omega_0)t}\, dt$$

This gives:

$$E(\nu) = 2\int_{0}^{\infty} e^{-\gamma u/(2(\omega - \omega_0))}\, e^{i\,u}\, \frac{du}{\omega - \omega_0}$$

Then we have:

$$E(\nu) = -\frac{2}{\omega - \omega_0}\int_{0}^{\infty} e^{-(\gamma u/(2(\omega - \omega_0)) - iu)}\, du$$

$$= -\frac{2}{\omega - \omega_0}\, \frac{1}{\frac{\gamma}{2(\omega - \omega_0)} - i}\, e^{-(\gamma/(2(\omega - \omega_0)) - i)u}\bigg|_{0}^{\infty} = \frac{2}{(\gamma/2) - i(\omega - \omega_0)}$$

(d) The line shape is given by $I(\nu)$, where the line intensity I is derived from

$$I(\nu) = |E(\nu)|^2 = \frac{4}{(\gamma/2)^2 + (\omega - \omega_0)^2}$$

We need to normalize $I(\nu)$: $\int_{-\infty}^{\infty} I(\nu)\, d\nu = 1$. This is done by determining k so that

$$1 = k\int_{-\infty}^{\infty} \frac{4}{(\gamma/2)^2 + (\omega - \omega_0)^2}\, d\nu = \frac{4k}{2\pi}\int_{-\infty}^{\infty} \frac{1}{(\gamma/2)^2 + (\omega - \omega_0)^2}\, d\omega$$

$$= \frac{2k}{\pi}\, \frac{2}{\gamma}\, \arctan\left(\frac{\omega - \omega_0}{\gamma/2}\right)\bigg|_{-\infty}^{\infty}$$

Thus, the result is $\frac{4k}{\gamma}$ and with

$$(\omega - \omega_0)^2 = 4\pi^2 (\Delta\nu)^2$$

So the line shape is described by

$$I(\nu) = \frac{\gamma}{(\gamma/2)^2 + 4\pi^2 (\Delta\nu)^2}$$

(e) Using the Doppler relation to convert from $f(v)$ to $f(\nu)$ results in

$$f(\nu) = \left(\frac{m}{2kT}\right)^{3/2} \exp(-\frac{m(\Delta\nu)^2 c^2}{2kT\nu_0^2})\,.$$

Normalizing as before to $f(\nu)$: $\int_{-\infty}^{\infty} f(\nu)d\nu = 1$ and substituting u^2 for the exponent implies

$$1 = \text{const} \times \left(\frac{m}{2kT}\right)^{3/2} \int_{-\infty}^{\infty} e^{-u^2}\, du \left(\sqrt{\frac{m}{2kT}}\frac{c}{\nu_0}\right)^{-1}$$

Solving for the integral ($\int_{-\infty}^{\infty} e^{-u^2}\, du = \sqrt{\pi}$) lets us determine the constant as const $= (2kTc)/(\sqrt{\pi}\nu_0 m)$. Thus, the line shape for thermal motion is given by

$$f(\nu) = \sqrt{\frac{m}{2\pi kT}}\,\frac{c}{\nu_0}\exp\left(-\frac{m(\Delta\nu)^2 c^2}{2kT\nu_0^2}\right).$$

(f) For a Lorentzian line shape, we determine the linewidth at the half power (HP) point by using $I_0(\nu = \nu_0) = 4/\gamma$. Thus, $I_{1/2} = 2/\gamma$ and $(\nu_{1/2} - \nu_0) = \Delta\nu(\text{HP}) = \gamma/(4\pi)$. This is half of the FWHM linewidth, $\Delta\nu_{1/2} = \gamma/(2\pi)$. In the same way, the HP linewidth for the Doppler profile is calculated as $\Delta\nu(\text{HP})^2 = (2\ln 2\, kT\nu_0^2)/(mc^2)$. Normalizing the lines with respect to each other requires their linewidths to be equal: $(2\ln 2\, kT\nu_0^2)/(mc^2) = \gamma^2/(16\pi^2)$. We now express the Doppler profile in terms of γ

$$f(\nu) = \sqrt{\frac{16\pi\,\ln 2}{\gamma^2}}\exp\left(-\ln 2\frac{16\pi^2}{\gamma^2}(\Delta\nu)^2\right).$$

the line wings are much more pronounced in the Lorentzian profile (fig. 28.1).
(g) For the Ly α line, we have $\omega_0 = 2.06 \times 10^{16}\,\text{s}^{-1}$. Approximating $\gamma = A$, we find for the 'natural' (FWHM) linewidth $\Delta\nu_{1/2} = A/(2\pi) = 8.59 \times 10^8\,\text{Hz}$. For the HI line ($\omega_0 = 8.92 \times 10^9\,\text{Hz}$), this linewidth is $2.06 \times 10^{-16}\,\text{Hz}$, extremely small. If pressure broadening due to collisions dominates, the lineshape is also Lorentzian,

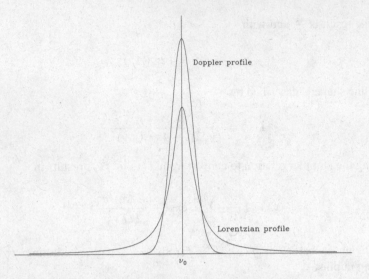

Fig. 28.1 Problem 3 (f): a comparison between a Lorentzian and a Doppler line profile

but the linewidth is given by $\pi \Delta v_{1/2p} = \langle \sigma v \rangle n$. Taking $\langle \sigma v \rangle = 10^{-10}\,\mathrm{cm}^3\,\mathrm{s}^{-1}$, we find:

For $n = 1\,\mathrm{cm}^{-3}$; $\Delta v_{1/2p} = 3.18 \times 10^{-11}\,\mathrm{Hz}$;
for $n = 10^5\,\mathrm{cm}^{-3}$; $\Delta v_{1/2p} = 3.18 \times 10^{-6}\,\mathrm{Hz}$;
for $n = 10^{19}\,\mathrm{cm}^{-3}$; $\Delta v_{1/2p} = 3.18 \times 10^{8}\,\mathrm{Hz}$.

In all cases, this line is far broader than the natural linewidth for HI, but we get a large linewidth only at very high densities. Thus, in the ISM, the linewidths of several $\mathrm{km\,s^{-1}}$ (several kHz) that are typically observed are usually dominated by thermal motion (Doppler broadening) and Gaussian shaped.

4.*(a) The energy is given simply by the sum of the kinetic and electrostatic energy

$$E = \frac{1}{2}mv^2 - \frac{e^2}{x} = \frac{p^2}{2m} - \frac{e^2}{x}\,.$$

(b) From the given relations, we obtain $x = \hbar/p$. Inserting this into the energy equation and differentiating yields for the minimum value of p

$$0 = \frac{\mathrm{d}E}{\mathrm{d}p} = \frac{p}{m} - \frac{e^2}{\hbar} \quad \Rightarrow \quad p_{\min} = \frac{e^2 m}{\hbar}\,.$$

Thus, $x_0 = \hbar^2/(e^2 m) = 5.29 \times 10^{-9}\,\mathrm{cm}$, which is the lowest Bohr orbit. The total energy of an electron in this orbit is $E_0 = -e^2/(2x_0) = -2.18 \times 10^{-11}\,\mathrm{erg}$ $= -13.56\,\mathrm{eV}$.

Assuming $x_n = x_0 n^2$, we have $x_1 = x_0$ and $x_2 = 4x_0 = 2.12 \times 10^{-8}$ cm. The energy of the lowest orbit is $E_1 = E_0$, and for the second lowest orbit we have $E_2 = -3.38$ eV. With $|E_1 - E_2| = 10.18$ eV $= h\nu$, we find for the frequency $\nu = 2.47 \times 10^{15}$ Hz, corresponding to a wavelength of 1215 Å. This is in the UV, at a wavelength about 33% longer than the Lyman α line (911 Å).

5. This is shown by inserting the constants into the given equation, dividing the result by 10^{27} (Hz $\rightarrow$ GHz) and multiplying by 10^{-18} (esu $\rightarrow$ 10^{-18} esu = Debye).

Chapter 29
Solutions for Chapter 13: Line Radiation of Neutral Hydrogen

1. Using the given Boltzmann relation and $T_s = T_K$, we find for N_u/N_l:

$T_K = 3$ K: 2.932
$T_K = 100$ K: 2.998
$T_K = 10^4$ K: 2.99998
$T_K = 10^6$ K: 3.00

The difference in relative population between 3 and 10^6 K is only 2.3%.

2. (a) Substituting $x^2 = (mV^2)/(2kT_K)$, we have, in one dimension,

$$\int_{-\infty}^{\infty} e^{-\frac{mV^2}{2kT_K}} dV = \int_{-\infty}^{\infty} \sqrt{\frac{2kT_K}{m}} e^{-x^2} dx = \sqrt{\frac{2kT_K}{m}} \sqrt{\pi} .$$

Thus, in three dimensions, a factor of $(m/2\pi kT_K)^{3/2}$ is needed for normalization.
(b) With the expression for V_{RMS} provided, we find

$$V_{RMS} = \sqrt{\int_{-\infty}^{\infty} v^2 f(v) dv} = \sqrt{3\frac{kT_K}{m}} .$$

Using $\Delta V_{1/2} = \sqrt{8\ln 2/3}\ V_{RMS}$, we find for atomic hydrogen

$$\Delta V_{1/2}[\text{km s}^{-1}] = \sqrt{\frac{8\ln 2\ kT_K}{m_H}} = 0.217\sqrt{T_K} .$$

Setting $\Delta V_{1/2} = 1$ km s^{-1} yields $T_K = 21.2$ K.
(c) Solving the result from part **(b)** for T_K gives $T_K = (m/(8\ln 2\ k))(\Delta V_{1/2})^2 = m_H/(8\ln 2\ k)(m/m_H)\ (\Delta V_{1/2})^2 = 21.2(m/m_H)\ (\Delta V_{1/2})^2$.

© Springer International Publishing AG, part of Springer Nature 2018
T. L. Wilson, S. Hüttemeister, *Tools of Radio Astronomy – Problems and Solutions*, Astronomy and Astrophysics Library,
https://doi.org/10.1007/978-3-319-90820-5_29

(d) First, we check the dimensions: $[P] = \text{dyne cm}^{-2} = \text{g cm s}^{-2}\text{cm}^{-2}$ $= \text{g s}^{-2}\text{cm}^{-1}$. This gives $\left[\sqrt{P/\rho}\right] = \sqrt{\text{g s}^{-2}\text{cm}^{-1}/\text{g cm}^{-3}} = \text{cm s}^{-1}$, which is indeed a velocity. For a perfect hydrogen gas, we find (n = number density): $c_0[\text{km s}^{-1}] = \sqrt{(nkT_K)/(nm_H)} = 9.08 \times 10^{-2}\sqrt{T_K}$. For a temperature of 21 K, this is $c_0 = 0.42\,\text{km s}^{-1}$.

3*. If the energy of the collision is too small (i.e. smaller than the energy separating the two levels, E_0), the upper state cannot be excited, so $\sigma = 0$. If the energy is very large ($E \gg E_0$), there is a problem satisfying both energy and momentum conservation, so σ is small. The maximum σ lies in between in this case, since no other levels are involved.

4*. With the result of Problem 5(d) and equation (12.2), the interaction energy is

$$W = \frac{4}{3}\frac{4Z^3}{In^3}\frac{\mu_n\mu_e}{2}[F(F+1) - I(I+1) - J(J+1)].$$

For HI ($I = 1/2; J = 1/2; F_u = 1; F_l = 0, Z = 1; n = 1$) this becomes

$$\Delta W(\text{HI}) = W_u - W_l = \frac{4}{3}\times 16\frac{\mu_{n,\text{HI}}\mu_e}{2} = \frac{32}{3}\mu_{n,\text{HI}}\mu_e.$$

In the same way, we get for DI ($I = 1; J = 1/2; F_u = 3/2; F_l = 1/2, Z = 1; n = 1$)

$$\Delta W(\text{DI}) = W_u - W_l = \frac{4}{3}12\frac{\mu_{n,\text{DI}}\mu_e}{2} = 8\mu_{n,\text{DI}}\mu_e$$

Inserting the values for the nuclear magnetic moments of HI and DI results in $\Delta W(\text{HI})/\Delta W(\text{DI}) = 4.341$. Thus, $\nu(= 0.230\nu(\text{HI})$, or, with $\nu(\text{HI}) = 1.42\,\text{GHz}$, $\nu(\text{DI}) = 327\,\text{MHz}$.

5. An analogous calculation to the previous problem results in

$$\Delta W(^3\text{He}^+) = \frac{4}{3}(-128)\frac{\mu_n\mu_e}{2} = -\frac{256}{3}\mu_n\mu_e.$$

Applying the nuclear magnetic moments gives for the ratio between HI and $^3\text{He}^+$: $\Delta W(\text{HI})/\Delta W(^3\text{He}^+) = 0.16$. The frequency of the $^3\text{He}^+$ hyperfine transition is $\nu(^3\text{He}^+) = 6.10\nu(\text{HI}) = 8.665\,\text{GHz}$.

6. For DI, a spin-1 system, we have: $A_{ul}(\text{DI}) = \frac{(\nu(\text{DI}))^3}{(\nu\text{HI})^3}\frac{4}{3}A_{ul}(\text{HI}) = 4.63 \times 10^{-17}\,\text{s}^{-1}$. In the same way, one finds for $^3\text{He}^+$:

$$A_{ul}(^3\text{He}^+) = \frac{(\nu(^3\text{He}^+))^3}{(\nu\text{HI}))^3}3A_{ul}(\text{HI}) = 1.95 \times 10^{12}\,\text{s}^{-1}.$$

These results agree with the ones given in Table 13.1, given in the statement of this problem and in 'Tools'.

7.*(a) For an estimate of the FWHP beamwidth, it is sufficient to assume that the beamwidth scales inversely with frequency. Thus, we have: $\Theta_b(\text{DI}) = \frac{\nu(\text{HI})}{\nu(\text{DI})}\Theta_b(\text{HI}) = 39'$. The peak continuum (main beam brightness) temperature of Cas A at the DI frequency is given by $T_{\text{MB,cont}} = (\lambda^2 S)/(2.65\,\Theta_o^2)$, with λ in cm and $\Theta_o = \sqrt{\Theta_b^2 + \Theta_s^2}$ in arc minutes (for Gaussian-shaped sources and beams, see Eq. (8.20) in 'Tools'). Since the source (5.5') is much smaller than the beam (39'), Θ_o is dominated by the beam size: $\Theta_o = 39.4'$. With the given spectral index of 0.7, the flux S at 92 cm is
$S(\text{DI}) = (\nu(\text{DI})/\nu(\text{HI}))^{0.7}\,S(\text{HI}) = 8384\,\text{Jy}$. This results in
$T_{\text{MB,cont}}(\text{DI}) = 17,150\,\text{K}$.
(b) The Doppler formula gives $\Delta\nu_1 = \nu_1\,(\Delta V/c)$. A velocity width of $2\,\text{km s}^{-1}$ thus corresponds to a frequency width of 2.18 kHz at $\nu_1 = 327\,\text{MHz}$.
(c) Simplifying the given formula by assuming $h\nu \ll kT_{\text{ex}}$, setting $T_{\text{ex}} = T_s$ and inserting numerical values, yields

$$N_1 = 1.95 \times 10^3 \frac{g_1\nu^2[\text{GHz}]}{g_u A_{\text{ul}}} T_s \int \tau\,dV\,.$$

Evaluating this formula for the case of DI gives $N_1(\text{DI}) = 2.25 \times 10^{18}\,T_s \int \tau\,dV$. The total column density is $N_{\text{tot}} = N_1 + N_u$. In a very good approximation, the ratio of the upper and lower level populations is given by the ratio of the statistical weights: $N_u/N_1 = 2$. The integral is (for a Gaussian line shape) equal to $\int \tau\,dV = 1.06\,\tau_{\text{peak}}\Delta V_{1/2}$. Therefore, we have for the total column density $N_{\text{tot}}(\text{DI}) = 3N_1 = 7.16 \times 10^{18}\,T_s\,\tau_{\text{peak}}\,\Delta V_{1/2}$.
(d) From 'Tools', Eq. (12.17), we have:

$$\kappa_\nu = \frac{c^2}{8\pi}\frac{1}{\nu_0^2}\frac{g_2}{g_1} n_1 A_{21}\left[1 - \exp\left(-\frac{h\nu_0}{kT}\right)\right]\varphi(\nu)$$

Replacing $\varphi(\nu) = \Delta\nu$ by ΔV requires use of the Doppler relation: $\frac{\Delta\nu}{\nu_0} = \frac{\Delta V}{c}$, where 'c' is the speed of light. In units of km s^{-1} and GHz, we have $\Delta\nu = 3 \times 10^{-4}/\nu(\text{GHz}) \times \Delta V$ and $\frac{h}{k} = 4.8 \times 10^{-2}\,\text{K GHz}^{-1}$. In addition, the integral over path length converts κ into τ, the optical depth, and n_1 into N_1, the column density. Then the relation is

$$\tau\Delta V = 1.07 \times 10^{-2} A_{21} \times N_1 \times \frac{g_2}{g_1}\left[1 - \exp\left(-\frac{h\nu_0}{kT}\right)\right]$$

Solving for column density, we obtain the relation given in the statement of this problem.

(e) To determine the antenna temperature of the DI line, its (apparent) optical depth is needed. This can be calculated from the known optical depth of the HI transition by first obtaining a relation like the one for the DI column density (part c) for HI: $N_1(\text{HI}) = 4.84 \times 10^{17}\ T_s\ \tau_{\text{peak}}\ \Delta V_{1/2}$. Using $N_u/N_1 = g_u/g_1 = 3$ yields for $N_{\text{tot}}(\text{HI}) = 4N_1 = 1.93 \times 10^{18}\ T_s\ \tau_{\text{peak}}\ \Delta V_{1/2}$. The optical depth is now given by the ratio of the column densities of DI and HI, which is equal to the D/H abundance ratio:

$$1.5 \times 10^{-5} = \frac{7.16 \times 10^{18}\ T_s\ \tau_{\text{peak}}(\text{DI})\ \Delta V_{1/2}}{1.93 \times 10^{18}\ T_s\ \tau_{\text{peak}}(\text{HI})\ \Delta V_{1/2}} = 3.71\ \frac{\tau_{\text{peak}}(\text{DI})}{\tau_{\text{peak}}(\text{HI})}\ .$$

With $\tau(\text{HI}) = 2.5$, this gives $\tau(\text{DI}) = 1.01 \times 10^{-5}$, and $T_{\text{line}}(\text{DI}) = 0.173\,\text{K}$ on a T_{MB} scale. With a beam efficiency of $\eta_b = 0.75$, the antenna temperature is $T_A^*(\text{DI}) = \eta_b T_{\text{MB}} = 0.13\,\text{K}$. The system temperature is entirely dominated by the strong continuum emission of Cas A: $T_{\text{sys}} = 100\,K + T_{\text{MB,cont}} = 17{,}242\,\text{K}$ on a T_{MB} scale, or 12,932 K on a T_A^* scale. Finally, the integration time is given by the radiometer formula (for On–Off observations)

$$\Delta T_{\text{RMS}} = \frac{2T_{\text{sys}}}{\sqrt{B\,t}} \quad \Rightarrow \quad t = \frac{4T_{\text{sys}}^2}{\Delta T_{\text{RMS}}^2\,B}\ ,$$

where B is the bandwidth in Hz (for 1/2 the FWHP width of the DI line, 1090 Hz, see part (b)). For a credible detection, we require a signal-to-noise ratio of 5σ, thus $\Delta T_{\text{RMS}} = 2.6 \times 10^{-2}\,\text{K}$. This results in an integration time of $9.08 \times 10^8\,\text{s}$ or approximately 29 years, which is an integration time beyond anything possible with a large radiotelescope. It is far longer than the—already very long—integration of 1253 hours (52 days) obtained by Heiles et al. 1993 (Ap. J. Suppl. **89**, 271), which resulted in a non-detection.

More recently Chengalur, Braun and Burton 1997 (A & A, 1997 318, L35) and Rogers, Dudevoir and Bania (2007 A. J. 133, 1625) have carried out searches in the outer galaxy. The first group reported a 3.9σ detection of the D I line with the Westerbork array while the second group found a 9σ detection, with a dipole array. The D I emission in the outer galaxy is more favorable since the continuum background is lower, so the system noise is less. Also the D I is easily destroyed by nuclear reactions, so should be more abundant in the outer galaxy where star formation is lower. The detection by Rogers et al. (2007) required the equivalent of many years of observations.

8. (a) The basic equation of radiative transfer that is valid here (in the Rayleigh–Jeans limit) is

$$T_L = T_{\text{cont}}e^{-\tau} + T_s(1 - e^{-\tau})\ ,$$

and

$$\Delta T_L = T_L - T_{cont} = (T_s - T_{cont})(1 - e^{-\tau}) .$$

The only contribution to the continuum temperature is the 2.73 K cosmic background radiation, thus $T_{cont} = T_{BG} = 2.73$ K. We can assume that $T_s = T_K$. Thus, the temperatures cancel and we will see no line at all.

(b) If the (main beam) continuum background temperature exceeds the spin temperature, $T_L < 0$, i.e. the line will appear in absorption. For $T_{BG,MB} = 3$ K and $\tau = 1$, we get $T_L = -0.17$ K.
(c) Now, T_K and thus T_s exceeds T_{BG} and $T_L > 0$, i.e. an emission line results. Again assuming $\tau = 1$, and $T_{BG} = 2.73$ K, we find $T_L = +0.48$ K.

9. We assume that all sources are Gaussian shaped. The continuum source fills a fraction of the radio telescope beam. This is given by:

$$\frac{\theta_0^2}{\theta_0^2 + \theta_a^2} \approx \frac{\Delta\Omega_0}{\Delta\Omega_a}$$

(with $\Delta\Omega$ denoting solid angles), since the beam size is taken to be much larger than all source sizes. In analogy, the fraction of the HI cloud filling the beam is $f^2\Delta\Omega_0/\Delta\Omega_a$. Only this fraction of the continuum source contributes to the absorption term in the relation for the line brightness temperature. Thus, the main beam brightness temperature of the line is given by

$$\Delta T_L = \frac{f^2\Delta\Omega_0}{\Delta\Omega_a}(T_{cl} - T_0)(1 - e^{-\tau}) .$$

Since f^2, $\Delta\Omega_0$ and $\Delta\Omega_a$ are all positive, the criterion to obtain absorption is simply $T_{cl} < T_0$, which is very likely, since the actual brightness temperature of, e.g., an HII region is high.
In many cases $T_{cl} \ll T_0$. If, in addition, the optical depth is low, the relation for ΔT_L can be rewritten as

$$\Delta T_L = \frac{f^2\Delta\Omega_0}{\Delta\Omega_a}(T_{cl} - T_0)(1 - e^{-\tau}) \approx -\frac{f^2\Delta\Omega_0}{\Delta\Omega_a}T_0\tau .$$

Thus, $|\Delta T_L/T_0| = (f^2\Delta\Omega_0/\Delta\Omega_a)\tau = f_{cl}\tau$ (with f_{cl} being the (area) beam filling factor of the cloud) is a good approximation in a number of realistic cases.

10. We assume that the larger HI cloud is in front of the continuum source; if it is not, the absorption term in the equation for the main beam will vanish. In this case, we will always see an emission line. Since both the HI- and the continuum source

are much smaller than the beam, their (area) beam filling factors are

$$\frac{\theta_{cl}^2}{\theta_{cl}^2 + \theta_a^2} \approx \frac{\Delta\Omega_{cl}}{\Delta\Omega_a} \quad \text{and} \quad \frac{\theta_0^2}{\theta_0^2 + \theta_a^2} \approx \frac{\Delta\Omega_0}{\Delta\Omega_a} \ .$$

The main beam brightness temperature is given by

$$\Delta T_L = \underbrace{(T_{cl} - T_0)(1 - e^{-\tau})\frac{\Delta\Omega_0}{\Delta\Omega_a}}_{\text{absorption if } T_{cl} < T_0} + \underbrace{T_{cl}(1 - e^{-\tau})(\frac{\Delta\Omega_{cl}}{\Delta\Omega_a} - \frac{\Delta\Omega_0}{\Delta\Omega_a})}_{\text{net emission}}$$

This gives:

$$\Delta T_L = -T_0\frac{\Delta\Omega_0}{\Delta\Omega_a}(1 - e^{-\tau}) + T_{cl}\frac{\Delta\Omega_{cl}}{\Delta\Omega_a}(1 - e^{-\tau}) \ .$$

Defining the 'main beam temperatures' $T_{0,MB} \equiv T_0(\Delta\Omega_0/\Delta\Omega_a)$ and $T_{cl,MB} \equiv (\Delta\Omega_{cl})/(\Delta\Omega_a)$, the main beam beam brightness temperature becomes

$$\Delta T_L = (T_{cl,MB} - T_{0,MB})(1 - e^{-\tau}).$$

The criterion for absorption is thus $T_{cl,MB} < T_{0,MB}$ (the temperature of the HI cloud, diluted by the beam, must be less than the beam diluted continuum temperature).

11. In this case, $\theta_0, \theta_{cl} \gg \theta_a$. Thus, all beam filling factors approach unity and the main beam brightness temperature is simply determined by

$$\Delta T_L = (T_{cl} - T_0)(1 - e^{-\tau})$$

as in Problem 11. Thus, an absorption line is seen if $T_{cl} < T_0$; the physical temperature of the continuum source must exceed the physical temperature of the line source, similar to the result obtained in Problem 13.

Chapter 30
Solutions for Chapter 14: Recombination Lines

1*. (a) The excitation parameter U is the radius of the HII region (in pc), multiplied by the 2/3 power of the electron density. This quantity characterizes an HII region; it is equivalent to $[3R_* N_{Lc}/\alpha_t]^{1/3}$ (R_*: radius of the exciting star, N_{Lc}: Lyman continuum photons per unit surface, α_t: this is the effective recombination probability). Thus, U relates the size and density of an HII region to parameters that depend *only* on the properties of the exciting star.

(b) $s_0 = U/N_e^{2/3}$ yields a radius of 0.146 pc.

(c) The emission measure of this HII region is EM$= 2.93 \times 10^7$ pc cm^{-6}.

(d) The mass of the HII region (assuming pure hydrogen) is 6.36×10^{33} g or $3.2\,M_\odot$.

2*. Since $U_{tot} = \sum_i U_i$, $2.8 \approx 3$ stars of spectral type B0 are needed to equal the number of ionizing photons from one O7 star.

3*. (a) The O7 star is $\sim$15.6 times more massive than the HII region that surrounds it. The ratio of stellar mass to the mass of the HII region remains roughly the same if an equivalent number of early B-type stars are considered instead of a single O star.

(b) The mass of an HII region with the same EM as in Problem 1 but $N_e = 3 \times 10^4$ cm^{-3} is $0.013\,M_\odot$. It has a radius that is nine times smaller than that in Problem 1 (0.016 pc), resulting in a volume that is smaller by a factor of 729, and a mass that is smaller by a factor of 243, i.e. for the same EM, higher density HII regions have smaller masses.

(c) For a density of $N_e = 3 \times 10^3$ cm^{-3}, $U = 68$ pc cm$^{2/3}$ results in a radius $s_0 = 0.327$ pc and a mass of $10.8\,M_\odot$. Thus, for the same U, lower density HII regions have larger masses.

4. The line-of-sight depth follows from $L = $ EM$/N_e^2$ as 0.04 pc. For the same EM, the RMS density is 2.7×10^3 cm^{-3}, and the clumping factor is $f = N_e / \langle N_e \rangle_{RMS} = 3.67$. It is a measure for the inhomogeneousness of the HII region.

© Springer International Publishing AG, part of Springer Nature 2018
T. L. Wilson, S. Hüttemeister, *Tools of Radio Astronomy – Problems and Solutions*, Astronomy and Astrophysics Library,
https://doi.org/10.1007/978-3-319-90820-5_30

5. Integrating over the Salpeter distribution yields $M_{tot} = 1.53N_0$ $\times M^{0.65}|_{M(\text{lower})}^{M(\text{upper})}$. With $M_{lower} = 0.08\,M_\odot$ and $M_{upper} = 50\,M_\odot$, this becomes $M_{tot} = 1.53N_0(12.71 - 0.19) = 19.16N_0$. Drawing the line at $18\,M_\odot$ (spectral type B0) we find $M_{tot}(< 18\,M_\odot) = 9.72N_0$ and $M_{tot}(> 18\,M_\odot) = 9.44N_0$.

Thus, the total mass in the lower mass stars (so-called later spectral types) is equal to the mass in higher mass stars (so-called B0 and earlier spectral type stars). However, the *number* of lower mass stars is much larger.

6*. (a) The sharp drop corresponds to the ionization potential of the atom, since it is assumed that one photon provides all the energy for the ionization. The energy corresponding to ν_0 is the ionization energy, $E_i = h\nu_0$. For H, this is 2.18×10^{-11} erg or 13.6 eV, for He, it is 3.96×10^{-11} erg or 24.6 eV.
(b) If only higher energy photons reach the outer part of an HII region and are absorbed, the kinetic temperature of the electrons released by these photons should be higher. Thus, the electron temperature $T_e = T_K$ should rise away from the center of an HII region.

7. Inserting the atomic masses into the given formula yields
$R_M(D) = 3.288947 \times 10^{15}$ Hz (deuterium) and
$R_M(^3He) = 3.289244 \times 10^{15}$ Hz (3-helium).

8. (a) To calculate the velocity difference between the ^{4}He- and ^{3}He-lines, we use

$$\Delta v = c\,\frac{\Delta \nu}{\nu_\infty} = c\,\frac{Z^2(\frac{1}{n^2} - \frac{1}{k^2})(R_M(^4He) - R_M(^3He))}{Z^2(\frac{1}{n^2} - \frac{1}{k^2})R_\infty}$$

$$= c\,\frac{R_M(^4He) - R_M(^3He)}{R_\infty}\ .$$

where R_∞ is the Rydberg constant for a nucleus of infinite mass. Inserting the appropriate constants yields $\Delta v = 13.37\,\mathrm{km\,s^{-1}}$. Thus, the lines overlap close to the half-power point, so the two profiles are blended, even if the linewidths are $20\,\mathrm{km\,s^{-1}}$.
(b*) Again assuming low optical depth, the line temperature of the ^{3}He transition will be 2×10^{-4} K. A detection is constituted by a signal-to-noise ratio of 5, i.e. the noise has to be 4×10^{-5} K. Using $t_{int} = (K^2 T_{sys}^2)/(B T_{RMS}^2)$ with $K = 2$

for position switching, T_{sys} the system noise, B the bandwidth and T_{RMS} the spectral noise to be reached, an observing time of 444 h or 18.5 days results (without taking into account telescope overhead). Even after this time, structure in the blended ^{4}He line might hide the ^{3}He line.

Table 30.1 Parameters of selected radio recombination lines (problem 9)

Transition	Exact frequency (MHz)	Approximate frequency (MHz)	Error (MHz)
100α	6478.659	6576.000	97.341
109α	5008.844	5077.878	69.035
126α	3248.656	3287.382	38.726
166α	1424.711	1437.598	12.887

9. (a) For $k = n + 1$ we have

$$\nu = Z^2 R_M \left(\frac{1}{n^2} - \frac{1}{(n+1)^2} \right) = Z^2 R_M \frac{2n+1}{n^2(n+1)^2} \approx Z^2 R_M \frac{2n}{n^4}$$

$$= \frac{2Z^2 R_M}{n^3} .$$

(b*) The exact and approximate frequencies along with the frequency error are listed in the table below (for $R_M = 3.288 \times 10^{15}$ Hz and $Z = 1$).
Even for the 166α transition, the error exceeds 10 MHz (Table 30.1).

10. The width of a Doppler-broadened (Gaussian-shaped) line is proportional to $1/\sqrt{M}$, with M the atomic mass. Thus, the formula valid for hydrogen has to modified to $\Delta V_{1/2} = \sqrt{0.04567 T_e/M}$ (neglecting micro turbulence). Accordingly, the linewidth for ^{4}He ($T_e = 10^4$ K, $M = 4$H) is 10.68 km s^{-1}, and the linewidth for ^{12}C ($T_e = 100$ K, $M = 12$H) is 0.616 km s^{-1}. At the peak of the ^{12}C line, the ^{4}He line has fallen to a level of 1.22×10^{-8} of its peak intensity (using $I(\Delta V) \sim \exp(-(\Delta V)^2/(2\sigma^2))$ and $\Delta V_{1/2} = 2.354\sigma$, i.e. Gaussian shaped lines). At the half-power point of the ^{12}C line, the contribution of the ^{4}He line is 1.83×10^{-8}. Thus, there is no visible overlap and the lines are clearly separated. Adding a turbulent width of $\nu_t = 20$ km s^{-1} gives $\Delta V_{1/2} = \sqrt{0.04567\, T_e/M + \nu_t^2} = 22.67$ km s^{-1}.
Now, the overlap (at the peak of the ^{12}C line) is at the 1.75×10^{-2} level, i.e. the lines are still clearly separated. However, the actual separation in velocity for Orion A is different, because ^{4}He and ^{12}C are not at the same ν_{LSR}. In the case of Orion A, the carbon line is closer to the helium line, from observations, since the carbon line arises in a partially ionized interface which may have a different radial velocity.

11.* (a) Using the approximation of Problem 9, we have $Z = 1$ for the recombination of helium and $Z = 2$ for the recombination of the remaining electron of singly ionized helium. The (approximate) frequency in the latter case is $\nu_k, n(\text{He}^+) = 8R_M/n^3$. To obtain the same frequency, the requirement for the principal quantum number is $n(\text{He}^+) = 1.587 n(\text{He})$.
(b) Substituting the Bohr radius into the expression for the dipole moment gives

$$\mu_{n+1,n} = \frac{1}{2} e \times 5.29 \times 10^{-9} \frac{n^2}{Z^2} .$$

Inserting this and the approximate frequency into the general formula for the A coefficient yields

$$A_{ul} = \frac{64\pi^4}{3hc^3} \, v^3 \, |\mu|^2 = 1.165 \times 10^{-2} \left(\frac{2Z^2 R_M}{n^3}\right)^3 \left(1.271 \times 10^{-18} \frac{n^2}{Z^2}\right)^2 .$$

The evaluation of all constants confirms the result

$$A_{ul} = 5.36 \times 10^9 \frac{Z^2}{n^5} .$$

Inserting this result into the general relation for the (velocity-integrated) intensity of an optically thin line yields

$$T_L \, \Delta v_{1/2} = 4.34 \times 10^{-4} \frac{g_u}{g_l} \frac{A_{ul}}{v^2[\text{GHz}]} \, N_l = 2.32 \times 10^6 \frac{g_u}{g_l} \frac{N_l}{v^2} \frac{Z^2}{n^5} .$$

We are considering transitions at the same frequency, and assume the same linewidth, thus:

$$T_L \sim \frac{g_u}{g_l} \frac{Z^2}{n^5} \, N_l .$$

Assuming $g_u/g_l \sim 1$ for high n, the line intensity ratio for He and He$^+$ is (using the result from part (a))

$$\frac{T(\text{He})}{T(\text{He}^+)} = \frac{[Z(\text{He})]^2 \, [n(\text{He})]^{-5}}{[Z(\text{He}^+)]^2 \, [n(\text{He}^+)]^{-5}} \frac{N_l(\text{He})}{N_l(\text{He}^+)} = 2.517 \frac{N_l(\text{He})}{N_l(\text{He}^+)} .$$

The level population N_l is proportional to principal quantum number squared, that is n^2, so the density of the atom or ion in question is:

$$\frac{T(\text{He})}{T(\text{He}^+)} = 2.517 \frac{[n(\text{He})]^2}{[n(\text{He}^+)]^2} \times \frac{N(\text{He})}{N(\text{He}^+)} = 2.517 \frac{[n(\text{He})]^2}{2.517 \, [n(\text{He})]^2} \times \frac{N(\text{He})}{N(\text{He}^+)}$$

$$= \frac{N(\text{He})}{N(\text{He}^+)} .$$

Thus, which line (at the same frequency) is stronger depends only on the ratios of the relative column densities.

12. With $\tau_L \ll 1$ the equation for the line intensity reduces to $T_L = T_e e^{-\tau_c} \tau_L$. Since $e^{-\tau_c} \ll 1$ for large τ_c, the line intensity is indeed reduced, possibly to the point where the line becomes invisible, in the presence of optically thick continuum radiation. If τ_c is assumed to be small, the T_e that is determined is given by $T_e(1) = T_L/\tau_L$. If τ_c is large, this changes to $T_e(2) = (T_L/\tau_L)e^{\tau_c}$. Since $e^{\tau_c} \gg 1$, it follows that $T_L(2) \gg T_L(1)$: the T_e determined for an unsuspected, large τ_c is (much) larger than what is determined if τ_c is small.

13*. Using the Boltzmann distribution, $g_n = 2n^2$ for the statistical weights, and the line frequency from Problem 9, we determine for the ratio of the level populations in case of LTE

$$\frac{N_{101}}{N_{100}}(\text{LTE}) = \frac{g_{101}}{g_{100}} \exp\left(-\frac{h\nu}{kT_e}\right) = \frac{20\,402}{20{,}000} \exp\left(-\frac{0.311}{10{,}000}\right) = 1.020068 \ .$$

Applying the given ratio of the b_n values yields

$$\frac{N_{101}}{N_{100}}(\text{actual}) = 1.0201 \, \exp\left(-\frac{0.311}{T_{\text{ex}}}\right)$$

$$\rightarrow \quad \exp\left(-\frac{0.311}{T_{\text{ex}}}\right) = 1.0010783 \ .$$

From this, the excitation temperature follows as

$$1.0777 \times 10^{-3} = -\frac{0.311}{T_{\text{ex}}} \quad \rightarrow \quad T_{\text{ex}} = -298 \,\text{K} \ .$$

Thus, the excitation temperature is negative, since the level populations are inverted. Setting $T_e = \infty$ and defining r_b as the ratio of the b_n, we get

$$\frac{N_{101}}{N_{100}}(\text{actual}) = r_b \, \frac{g_{101}}{g_{100}}$$

$$= \frac{g_{101}}{g_{100}} \exp\left(-\frac{0.311}{T_{\text{ex}}}\right) \quad \rightarrow \quad r_b = \exp\left(-\frac{0.311}{T_{\text{ex}}}\right) \ .$$

If we, for example, require $T_{\text{ex}} > 10^4$ K for superthermal populations, the limiting r_b is

$$r_b = \exp\left(-\frac{0.311}{10000}\right) = 0.999969 \ .$$

Higher T_{ex} requires r_b to be even closer to (but smaller than) unity.

14. Following the same scheme as in the first part of Problem 13, we find

$$\frac{N_{41}}{N_{40}}(\text{actual}) = 1.005 \, \frac{g_{41}}{g_{40}} \exp\left(-\frac{4.94}{10000}\right) = \frac{g_{41}}{g_{40}} \exp\left(-\frac{4.94}{T_{\text{ex}}}\right)$$

$$\rightarrow \quad 1.004504 = \exp\left(-\frac{4.94}{T_{\text{ex}}}\right) \ .$$

From this, T_{ex} follows as $T_{ex} = -4.94/0.0044935 = -1099\,\text{K}$. Note that higher negative temperatures are an indication of level populations that are closer to LTE, that is, are less inverted.

15* Evaluating the equation given in the problem yields $\beta = -284$. The measure of the influence of NLTE effects follows to $T_L/T_L^* = 0.9692(1 + 142\tau_c)$. Clearly, it rises as the continuum radiation becomes more optically thick.

16*. First, we determine the optical depth of the continuum radiation by using Eq. (10.37) in 'Tools' and the statement of problem 3 in chapter 10. This equation is: $\tau_c = 8.235 \times 10^{-2}\,T_e^{-1.35}\,\nu^{-2.1}\,\text{EM} = 3.12 \times 10^{-5}$. The continuum radiation is very optically thin, thus the continuum radiation has the strength $T_c = T_e\tau_c = 3.12 \times 10^{-3}\,\text{K}$.

The strength of the line radiation is determined from Eq. (14.28) in 'Tools'; which is

$$T_L = 1.92 \times 10^3\,T_e^{-3/2}\,\text{EM}\,\Delta\nu^{-1} = 0.16\,\text{K}\,.$$

The line-to-continuum ratio is 51, i.e. large: small values for T_e give large line-to-continuum ratios.

To estimate the importance of non-LTE effects, we calculate, following Problem 17, $T_L/T_L^* = 0.75 \times (1 + 3.5\tau_c) = 0.75$. Here, T_L^* denotes the LTE line temperature. Thus, the NLTE temperature is $T_L = 0.75T_L^* = 0.12\,\text{K}$. That is, the difference between LTE and NLTE is not very large.

17. The fundamental equation of radiative transfer (Eq. (1.9) in 'Tools') gives $dI_\nu/ds = -\kappa_\nu I_\nu + \varepsilon_\nu$. If the source function S (equal to ε_ν) can be neglected compared to the background, this reduces to $dI_\nu/ds = -\kappa_\nu I_\nu$, or, equivalently, $dI_\nu/d\tau = I_\nu$. Thus, $I_\nu = I_0 e^\tau$. The optical depth is defined as $\tau_\nu = \kappa_\nu l$. Inserting Eq. (14.38) from 'Tools' gives $\tau_\nu = \kappa_\nu^* b\beta L$, so $I_\nu = I_0 e^{\kappa_\nu^* b\beta l}$, which is the relation that was to be shown. All quantities marked by $*$ denote the LTE case.

The measured line intensity is given by $\Delta I_L = I_L(l) - I_0 = I_0(e^{\tau_L} - 1)$. Using temperatures, this is $T_L = T_{BG}(e^{\tau_L} - 1) = T_{BG}(e^{\kappa_L^* b\beta l} - 1)$. We use the result from Problem 19 to obtain the optical depth of the line in the LTE case: $\tau_L^* = T_L/T_e = 1.6 \times 10^{-3} = \kappa_L^* l$. This yields $\tau_L = \kappa_L l = \kappa_L^* b\beta l = 8.4 \times 10^{-3}$. Finally, the observed NLTE line temperature is $T_L = 2500(e^{\tau_L} - 1) = 21.1\,\text{K}$.

18. (a) For a collision rate given by $t^{-1} = n\sigma v$, where n is the volume density (of hydrogen, the collision partner), v is estimated from the linewidth in Fig. 14.2 in 'Tools' as $4\,\text{km s}^{-1}$, and take σ as the collision cross section ($\sigma = \pi a_n^2$, a_n the Bohr radius). Inserting the numbers results in

$n = 300$: $t_c^{-1} = 10^3 \times 7.1 \times 10^{-7} \times 4 \times 10^4 = 28\,\text{s}^{-1}$.

$n = 100$: $t_c^{-1} = 10^3 \times 8.8 \times 10^{-9} \times 4 \times 10^4 = 0.35\,\text{s}^{-1}$.

The radiative decay is given by $A_{ul} = 5.36 \times 10^9 \, n^{-5}$ (for $Z = 1$, which is also correct for 'hydrogen-like' carbon lines). Thus, we get
$n = 300$: $t_r^{-1} = 2.21 \times 10^{-3} \, \text{s}^{-1}$.
$n = 100$: $t_r^{-1} = 0.54 \, \text{s}^{-1}$.
Collisions clearly dominate at $n = 150$ (then the population is close to LTE), while radiative decay starts to have higher rates at $n \approx 100$.
(b) Radiative transfer gives for a line temperature, ignoring filling factors, (e.g. Eq. (12.20) in 'Tools') $T_L = (T_{ex} - T_{BG})(1 - e^{-\tau})$. The second factor is always positive. Thus, to see any line in emission, T_{ex} has to exceed T_{BG}, or be negative (maser), as the observer correctly states for the C166α line.
(c) The equation is obtained as follows: the Doppler relation is used to convert the units of linewidth from kHz to km s^{-1}, as

$$\left(\frac{\Delta v}{\text{kHz}}\right) = 3.33 \left(\frac{v_0}{\text{GHz}}\right) \Delta V_{1/2} .$$

Then

$$T_L = 1.92 \times 10^3 \, T_e^{3/2} \, (\text{EM}) \times \left(\frac{\Delta v}{\text{kHz}}\right)^{-1}$$

becomes

$$T_L = 576 \, T_e^{3/2} \, (\text{EM}) \left(\frac{v_0}{\text{GHz}} \times \frac{\Delta V_{1/2}}{\text{km s}^{-1}}\right)^{-1} .$$

(d)* Using the method in Problem 9, chap. 9, we obtain $T(\text{Cas A}) = 9.9 \times 10^3$ K at 1.425 GHz. For the C$^+$ line region, from the values given in the problem, $n(\text{C}^+) = 1.2 \, \text{cm}^{-3}$, so the emission measure, EM, is 0.42 cm^{-6} pc. The most crucial input for an estimate of the peak line temperature is the value of T_e. If this emission is not caused by a maser amplification of the background continuum radiation, T_e must be > 0. In fact the value must exceed the maximum continuum brightness temperature $\sim 9.9 \times 10^3$ K, in order to have a C166α line is in emission. The predicted integrated line intensity is 1.7×10^{-4} K km s^{-1}. This is far smaller than the observed value. Thus, large non-LTE effects, namely line masering, must be present.

Chapter 31
Solutions for Chapter 15: Overview of Molecular Basics

1. (a) For an ideal gas, we have $PV = NkT$. With $N/V = n$, the pressure becomes $P = nkT$. For standard laboratory conditions, the density is: $n = 2.65 \times 10^{19} \, \text{cm}^{-3}$. For a molecular cloud with pressure $1.32 \times 10^{-9} \, \text{dyne cm}^{-2}$, $n = 9.56 \times 10^5 \, \text{cm}^{-3}$.
(b) Laboratory conditions: $\lambda = 1/(\sigma n) = 3.77 \times 10^{-4} \, \text{cm}$; $\tau = 1/(\sigma n v) = 1.26 \times 10^{-8} \, \text{s}$; molecular cloud: $\lambda = 1.05 \times 10^{10} \, \text{cm}$; $\tau = 5.23 \times 10^5 \, \text{s}$.
(c) Under laboratory conditions, there are 7.9×10^{12} collisions before a decay, in a molecular cloud there are 0.19 collisions before a decay. In the first case, collisions dominate over radiative decay, while in the second case, radiative decay is slightly faster.
(d) For the molecular cloud, the number of H atoms is twice the number density of molecules so $\lambda = 10^{15} \, \text{cm}$, while for the laboratory, $\lambda = 38 \, \text{cm}$. Quite a remarkable difference!

2. (a) For $m = 28 m_{\text{H}}$, we have for the thermal linewidth: $\Delta V_t = 4 \times 10 \times 10^{-2} \sqrt{T_K}$.

$T_K = 10 \, \text{K}$: $\Delta V_t = 0.13 \, \text{km s}^{-1}$.
$T_K = 100 \, \text{K}$: $\Delta V_t = 0.41 \, \text{km s}^{-1}$.
$T_K = 200 \, \text{K}$: $\Delta V_t = 0.58 \, \text{km s}^{-1}$.

(b) With $\Delta V_{\text{turb}} = \sqrt{\Delta V_{1/2} - \Delta V_t}$, $\Delta V_{\text{turb}} = 2.997 \, \text{km s}^{-1}$, i.e. turbulence is the dominating line broadening process.

3. Inserting the parameters (of the CO molecule) into the given equation yields $A_{ul} = 5.948 \times 10^{-8} \, \text{s}^{-1}$.

4. Using $n^* = \frac{A_{ul}}{\langle \sigma v \rangle} = 10^{10} A_{ul}$, we find:

$n^* = 1.8 \times 10^4 \, \text{cm}^{-3}$ for CS $J = 1 \to 0$.
$n^* = 2.2 \times 10^5 \, \text{cm}^{-3}$ for CS $J = 2 \to 1$.
$n^* = 740 \, \text{cm}^{-3}$ for CO $J = 1 \to 0$.

© Springer International Publishing AG, part of Springer Nature 2018
T. L. Wilson, S. Hüttemeister, *Tools of Radio Astronomy – Problems and Solutions*, Astronomy and Astrophysics Library,
https://doi.org/10.1007/978-3-319-90820-5_31

5. (a) For a rigid rotor (centrifugal stretching constant $D = 0$), the frequency of a transition is given by $\nu = 2B_e(J + 1)$, where the rotational constant $B_e = \hbar/(4\pi \Theta_e)$. This gives $B_e = 4.18 \times 10^{10}$ Hz. The resulting frequencies and energies above ground are

$J = 1 \to 0$: 83.6 GHz; energy of level $J = 1$: 4.01 K.
$J = 2 \to 1$: 167.1 GHz; energy of level $J = 2$: 12.03 K.
$J = 3 \to 2$: 250.7 GHz; energy of level $J = 3$: 24.06 K.
$J = 4 \to 3$: 334.4 GHz; energy of level $J = 4$: 40.10 K.

(b) For HD, we find $B_e = 1.348 \times 10^{12}$ Hz. This gives

$J = 1 \to 0$: 2696 GHz (or 111 μ) ; energy of level $J = 1$: 129 K.
$J = 2 \to 1$: 5392 GHz (or 56 μ); energy of level $J = 2$: 388 K.
$J = 3 \to 2$: 8088 GHz (or 37 μ); energy of level $J = 3$: 775 K.
$J = 4 \to 3$: 1.08×10^4 GHz (or 28 μ); energy of level $J = 4$: 1291 K.

(c) $J = 1 \to 0 : A_{ul} = 7.61 \times 10^{-10}\,\text{s}^{-1}$; critical density $n^* = 7.6\,\text{cm}^{-3}$.
$J = 2 \to 1 : A_{ul} = 7.30 \times 10^{-9}\,\text{s}^{-1}$; critical density $n^* = 73\,\text{cm}^{-3}$.

6. To determine the frequency, we use

$$\nu = \frac{E(J + 1) - E(J)}{h}$$
$$= B_e((J + 1)(J + 2) - J(J + 1)) - D((J + 1)^2(J + 2)^2 - J^2(J + 1)^2).$$

This reduces to $\nu = 2B_e(J + 1) - 4D(J + 1)^3$, which is Eq. (14.26) in 'Tools'. We find for the $^{12}\text{C}^{16}\text{O}$ molecule

$J = 1$: frequency: 115.273 GHz; energy: 7.638×10^{-16} erg or 5.53 K.
$J = 2$: frequency: 230.538 GHz; energy: 2.291×10^{-15} erg or 16.59 K.
$J = 3$: frequency: 345.796 GHz; energy: 4.583×10^{-15} erg or 33.19 K.
$J = 4$: frequency: 461.040 GHz; energy: 7.637×10^{-15} erg or 55.31 K.
$J = 5$: frequency: 576.267 GHz; energy: 1.146×10^{-14} erg or 82.98 K.

7. Using the same formulae as in Problem 6 of this chapter, we find

$J = 0$: energy: 0 erg
$J = 1$: frequency: 338.125 MHz; energy: 2.240×10^{-18} erg or 0.016 K.

To find an appropriate J for $\nu \approx 20$ GHz, we neglect the second term in the formula for the frequency (which contributes only ~2% even for $J \approx 100$), and set 20 GHz $= 2B_e(J + 1)$. This gives $J = 59$. We confirm our result by calculating the exact frequency and find: $\nu(J = 59) = 20.0802$ GHz.

8. Differentiating equation (14.49) gives

$$\frac{d(n(J))}{dJ} = \frac{2n_{\text{tot}}}{Z}e^{-hB_e J(J+1)/kT} - \frac{n_{\text{tot}}}{Z}e^{-hB_e J(J+1)/kT}\frac{hB_e(2J+1)}{kT}$$

$$= \frac{n_{\text{tot}}}{Z}e^{-hB_e J(J+1)/kT}\left(2 - \frac{hB_e(2J+1)^2}{kT}\right).$$

Setting this to 0 to find the maximum yields

$$0 = 2 - \frac{hB_e(2J+1)^2}{kT} \Rightarrow J_{\text{max}} = \sqrt{\frac{kT}{2hB_e}} - \frac{1}{2} = 3.227\sqrt{\frac{T}{B_e(\text{GHz})}} - \frac{1}{2}.$$

Applying this relation to selected diatomic molecules, we find:

CO; 10 K: $J_{\text{max}} = 0.84$ ($J = 1$ level); 100 K: $J_{\text{max}} = 3.75$ ($J = 4$ level).
CS; 10 K: $J_{\text{max}} = 1.56$ ($J = 2$ level); 100 K: $J_{\text{max}} = 6.01$ ($J = 6$ level).
HC$_{11}$N; 10 K: $J_{\text{max}} = 24.32$ ($J = 24$ level).

In the latter two cases, the required densities for LTE may be too large to be fulfilled.

9. (a) This is shown most simply by inserting the given quantities. We use $1 - e^{-0.048v/T_{\text{ex}}} \approx 0.048\, v/T_{\text{ex}}$ and $\int \tau\, dv = \tau\, \Delta v$.
(b) Since $T_{\text{MB}} = T_{\text{ex}}\tau \sim vn(J)$ and $v \sim J$, J_{max} will be higher.

10. We use the result established in Problem 9(a) and a level population following the Boltzmann relation. Applying $g_0 = 1$; $g_1 = 3$; $v_{2-1} = 2v_{1-0}$; $T_L = T_{\text{ex}}\tau$ and $(1/3)e^{-E(K)/T_{\text{ex}}} = 1/3$ for high T_{ex} results in

$$\frac{N_0}{N_1} = \frac{1}{3e^{-E(K)/T_{\text{ex}}}} = \frac{4T_{1-0}}{3T_{2-1}} \Rightarrow T_{2-1} = 4T_{1-0}.$$

For high optical depth, $T_L = T_{\text{ex}}$, so for a constant T_{ex}, the line intensities are equal: $T_{2-1} = T_{1-0}$.

11. (a) The partition function for ammonia is: $Z = 4.118 \times 10^{-2}\, T^{3/2}$. This gives: $T = 50\,\text{K} : Z = 14.6$; $T = 100\,\text{K} : Z = 41.2$; $T = 200\,\text{K} : Z = 116.5$; $T = 300\,\text{K} : Z = 214.0$. The approximation is valid as long as the temperature is large compared to the spacing of the rotational energy levels; for the ammonia molecule, so this must be $T \gg 14\,\text{K}$.
(b) Inserting the relation for Z into the given formula, taking into account that there are 2 $J = (3, 3)$ levels, results in

$$\frac{n_{\text{tot}}}{n_{3,3}} = \frac{Z}{2(2J+1)e^{-120/T}} = 2.941 \times 10^{-3}\, T^{3/2}\, e^{120/T}.$$

The ratios are: $T = 50\,\text{K}: 11.46$; $T = 100\,\text{K}: 9.76$; $T = 200\,\text{K}: 15.15$; $T = 300\,\text{K}: 22.80$.

(c) The given formula can be rewritten as

$$\frac{n_{\text{tot}}}{n_{3,3}} = \frac{Z}{2(2J+1)e^{-124.5/T}} = \frac{n_{0,0}}{14e^{-124.5/T}} + \frac{n_{1,1}}{14e^{-124.5/T}}$$

$$+ \frac{n_{2,2}}{14e^{-124.5/T}} + 1 + \frac{n_{4,4}}{14e^{-124.5/T}} + \cdots .$$

We insert $n(J = K) = (2J + 1) \exp((h(BJ(J+1) + (C - B)J^2)/(kT)))$, keeping in mind that for the $(J, K) = (3, 3)$ and $(9,9)$ transitions the factor $(2J+1)$ has to be replaced by $2(2J+1)$. Taking into account metastable levels with $J \leq 10$, we find:

$T = 50\,\text{K}: n_{\text{tot}}/n_{3,3} = 4.82;$
$T = 100\,\text{K}: n_{\text{tot}}/n_{3,3} = 3.05;$
$T = 200\,\text{K}: n_{\text{tot}}/n_{3,3} = 3.45;$
$T = 300\,\text{K}: n_{\text{tot}}/n_{3,3} = 4.23.$

Even for $T = 300\,\text{K}$, the (10,10) level contributes only 0.069 to the sum, so the approximation is justified.

12. (a) The symmetry of the total wavefunction is the product of the symmetries of the spin wavefunction times that of the spatial wavefunction. For formaldehyde or deuterated formaldehyde, the dipole moment is along the 'a' axis of the molecule. The quantum numbers of the rotation about the 'b' and 'c' axes determine the symmetry of the spatial wavefunction. Transitions between rotational states are characterized by $K_a K_c$. Since the deuterium spins are unity, these are bosons, so an exchange of these spin wave functions is symmetric. Since the total wavefunction must be symmetric, the para-D_2O (i.e. anti-parallel nuclear spins), these must be multiplied by an anti-symmetric spatial wavefunction. Since the dipole moment is along the 'a' direction, it must be symmetric, so for this species, transitions are between states with $K_a K_c$ even-even and/or odd-odd odd-even. For ortho-D_2CO (i.e. parallel nuclear spins), these must be multiplied by an symmetric spatial wavefunction. Since the dipole moment is along the 'a' direction, it must be symmetric, so for this species, transitions are between states with $K_a K_c$ * odd-odd and/or even-even (see, e. g. Butner et al. 2007 ApJ 659, L137).

Chapter 32
Solutions for Chapter 16: Molecules in Interstellar Space

1. (a) We use

$$\Delta v = \frac{\Delta E}{h} = \frac{E(J+1) - E(J)}{h} = 2B(J+1) + (C-B)K^2(K+1)^2$$

$$= 596\,\text{GHz}\,(J+1) - 109\,\text{GHz}\,K^2(K+1)^2 .$$

For the $(1,0) \rightarrow (0,0)$ transition, this results in $v = 596\,\text{GHz}$, and for the $(2,0) \rightarrow (1,0)$ transition, we get $v = 1192\,\text{GHz}$.
Inserting into the formula for A_{ul} given in Problem 3, the Einstein A coefficients are: $A((1,0) - (0,0)) = 1.78 \times 10^{-3}\,\text{s}^{-1}$ and $A((2,0) - (1,0)) = 1.71 \times 10^{-2}\,\text{s}^{-1}$. Compared to the inversion transitions with $A \approx 10^{-7}\,\text{s}^{-1}$, the rotational transitions are fast and the populations decay to the ground state quickly.

2. (a) Solving the Boltzmann equation for T_{ex} gives $T_{ex} = -1.14/\ln(n_u/n_l)$.
We obtain:
(b) Inserting into the formula from Problem 13 and solving for τ gives: $\tau_{1-0} = -6.98 \times 10^{-16}\,N(J=0)$.
(c) Inserting into the equation for T_{MB} gives $T_{MB} = -100\tau = 6.98 \times 10^{-14}\,N_0$. Thus, a measurement of T_{MB} allows the determination of N_0. For example, for $T_{MB} = 1\,K$, $N_0 = 1.43 \times 10^{13}\,\text{cm}^{-2}$. Since the background is not amplified in this case, optically thin masers indeed 'do not mase'.
(d) The condition $T_{BG} \gg T_{ex}$ lets us write: $T_{MB} = T_{BG}(e^{-\tau} - 1)$. Since τ is negative, the exponent is positive. For the optically thin case, there is amplification, for high optical depth, the background is very strongly amplified, since $(e^{-\tau} - 1) \gg 1$ (Table 32.1).

© Springer International Publishing AG, part of Springer Nature 2018
T. L. Wilson, S. Hüttemeister, *Tools of Radio Astronomy – Problems and Solutions*, Astronomy and Astrophysics Library,
https://doi.org/10.1007/978-3-319-90820-5_32

Table 32.1 Excitation
temperatures as a function of
ratios of populations
(problem 2)

n_u/n_l	T_{ex}
0.5	1.64
0.6	2.23
0.7	3.20
0.8	5.11
0.9	10.82
1.0	∞
1.1	−11.96
1.2	−6.25
1.3	−4.35
1.4	−3.39
1.5	−2.81

3. The conditions $A \gg C$ and $(A_{ji})/(3C_{ji}\tau_{ij})$ require $\tau \gg 1$. Applying this and $T_K/T_0 \gg 1$, the given equation simplifies to

$$\frac{T}{T_0} = \frac{1}{\ln\left(1 + (A_{ji})/(3C_{ji}\tau_{ij})\right)}.$$

For example, $(A_{ji})/(3C_{ji}\tau_{ij}) = 0.1$ results in $T_L = 10.5\, T_0$, so, even though the transition is subthermal (radiation dominates collisions), the line intensity is high, due to line trapping caused by the high optical depth, even though the gas density may be low.

4. The same equation used in Problem 3 holds. Since $(A_{ji})/(3C_{ji}\tau_{ij}) \ll 1$, this can be further reduced (by approximating the logarithm: $\ln(1 + x) \approx x$) to $T_L = 3C\tau/A\, T_0$. The condition $C \gg A$ indicates that collisions dominate and the gas is thermalized. The line intensity is high (e.g. $T_L = 10\, T_0$ for $(A_{ji})/(3C_{ji}\tau_{ij}) = 0.1$), but now also the density is high enough to allow collisions to be the most important process.

5*. **(a)** Let V be the uniform expansion velocity, $v_\perp$ its line-of-sight component, i.e. the radial velocity. For spatial coordinates, we choose p and z, with z aligned with $v_\perp$ and p the projection of r on the plane of the sky, so that for the radius r we have $r^2 = z^2 + p^2$. Using $v_\perp = V(z/r)$, we can express z as

$$z = \frac{v_\perp}{V}r = \frac{v_\perp}{V}\sqrt{z^2 + p^2} \;\Rightarrow\; z^2 = \frac{p^2(\frac{v_\perp}{V})^2}{1 - (\frac{v_\perp}{V})^2} \;\Rightarrow\; z = \frac{p\left(\frac{v_\perp}{V}\right)}{\sqrt{1 - (\frac{v_\perp}{V})^2}}.$$

Differentiating gives

$$\frac{dz}{dv_\perp} = \frac{\frac{p}{V}}{\sqrt{1 - (\frac{v_\perp}{V})^2}} - \frac{\frac{1}{2}p\left(\frac{v_\perp}{V}\right)\left(\frac{2v_\perp}{V}\right)}{\left(1 - (\frac{v_\perp}{V})^2\right)^{3/2}} = \frac{\frac{p}{V}}{\left(1 - (\frac{v_\perp}{V})^2\right)^{3/2}}.$$

We define a 'correlation length' $\delta z(p, v_\perp)$ over which the radial velocity is $v_\perp$, with ΔV the local linewidth due to turbulence and thermal broadening, which is small compared to the expansion velocity V

$$\delta z = \frac{\Delta V}{dv_\perp/dz} = \frac{p \, \Delta V}{V} \left(1 - \left(\frac{v_\perp}{V}\right)^2\right)^{-3/2}.$$

(b) To express τ as a function of p and $(v_\perp/V)$, we use the result of Problem 13 as the optical depth (per unit distance) and get

$$\tau(p, v_\perp) = \left(\frac{N_J(r)}{T_{\rm ex}} \frac{\mu_0^2 v(J+1)}{1.67 \times 10^{14} \, (2J+1)}\right) \frac{1}{\Delta V} \delta z$$

The result is

$$\tau(p, v_\perp) = C \frac{N_J(r)}{\Delta V} \delta z$$

$$= C \frac{N_J(r) p}{V} \frac{1}{(1 - (\frac{v_\perp}{V})^2)^{3/2}}.$$

(c) Here, the assumption of $\Delta V \ll V$ is important, since then all molecules contributing to $\tau(p, v_\perp)$ are at the same radius. The line intensity $T(v_\perp)$ (i.e. the line profile) is given by

$$T(v_\perp) = 2\pi \int_0^{p(v_\perp)} T_0(p, v_\perp)(1 - e^{-\tau(p,v_\perp)}) \, p \, dp.$$

(d) In the *optically thin case*, this becomes
$T(v_\perp) = 2\pi \int_0^{p(v_\perp)} T_0(p, v_\perp) \tau(p, v_\perp) \, p \, dp$. Changing the integration variable to r, we use $p \, dp = (1 - (v_\perp/V)^2) r \, dr$ and obtain

$$T(v_\perp) = 2\pi \int_0^{R_0} T_0(r) C \frac{N_J(r) p}{V} \frac{1}{(1 - (\frac{v_\perp}{V})^2)^{3/2}} (1 - (v_\perp/V)^2) r \, dr$$

$$= 2\pi \frac{C}{V} \int_0^{R_0} T_0(r) N_J(r) r^2 dr.$$

Because the integral is independent of $v_\perp$, the line profile will be flat out to the maximum velocity.

In the *optically thick case*, the profile function does not depend on τ and reduces to

$$T(v_\perp) = 2\pi \int_0^{p(v_\perp)} T_0(p, v_\perp) \, p \, dp = 2\pi \int_0^{R_0} T_0(r) \left(1 - \left(\frac{v_\perp}{V}\right)^2\right) r \, dr$$

$$= 2\pi \left(1 - \left(\frac{v_\perp}{V}\right)^2\right) \int_0^{R_0} T_0(r) \, r \, dr.$$

This is a parabolic line shape. In both cases, this analysis is only valid if the circumstellar envelope is much smaller than the beam of the telescope (see Morris 1975, ApJ 197, 603).

6*. (a) If the true length of the outflow is l and the true width w, the volume is given by $V = \pi(w/2)^2\, l = (1/4)\pi l w^2$. For a constant density $n(H_2)$, the mass (neglecting helium and heavier elements) is $M = nV = (1/4)\pi\, n(H_2)\, l w^2$.
(b) The age (assuming expansion at a constant rate) is given by age $= l/(2v)$ (v: true velocity). Using $v = v_0/\cos i$ and $l = l_0/\sin i$, we get age $= l_0/(2v_0 \tan i)$, where v_0 and l_0 are the observed quantities.
(c) The total kinetic energy is $E = (1/2)Mv^2 = (1/2)Mv_0^2/\cos^2 i$.
(d) For $\dot{E}$ we find
$E/\text{age} = \left[(Mv_0^2)/(2\cos^2 i)\,(2v_0 \sin i)/(l_0 \cos i)\right] = (Mv_0^3 \sin i)/(l_0 \cos^3 i)$.

7*. Using the equations from the statements of problems 3, 4 and 9 of chapter 15, we find

$J = 30 \to 29$: $\nu = 3.438 \times 10^3$ GHz; $\lambda = 0.087$ mm; $A_{ul} = 2.81 \times 10^{-3}$; $n^* = 2.81 \times 10^7$ cm^{-3}.
$J = 16 \to 15$: $\nu = 1.841 \times 10^3$ GHz; $\lambda = 0.163$ mm; $A_{ul} = 2.22 \times 10^{-3}$; $n^* = 2.22 \times 10^6$ cm^{-3}.
$J = 6 \to 5$: $\nu = 691.47$ GHz; $\lambda = 0.434$ mm; $A_{ul} = 1.86 \times 10^{-5}$; $n^* = 1.86 \times 10^5$ cm^{-3}.

For comparison: $J = 2 \to 1$: $\nu = 230.542$ GHz; $\lambda = 1.3$ mm; $A_{ul} = 6.9 \times 10^{-7}$; $n^* = 6.9 \times 10^3$ cm^{-3}.
(b) To determine the energies, the formula given in Problem 10 is used: $J = 30$: $E = 2566$ K; $J = 16$: $E = 752$ K; $J = 6$: $E = 116$ K. Assuming a Boltzmann distribution, we have $N(30)/N(6) = (61/13)e^{-2450/2000} = 1.38$. Since $C \approx 10^{-4}$ cm^{-3}, the $J = 30 \to 29$ transition will be subthermal, while the $J = 6 \to 5$ transition is thermalized. Thus, non-LTE (e.g. LVG) calculations are needed to determine the population ratio.

8. With $t_{\text{gas-grain}} = \frac{10^{22}}{n_H V_{1/2}}$ and for $T_k = 10$ K, we get

$$t_{\text{gas-grain}} = \frac{7.69 \times 10^{17}\,\text{s}}{n_H} = \frac{2.4 \times 10^{10}\,\text{yr}}{n_H} = \frac{1.2 \times 10^{10}\,\text{yr}}{n_{H_2}}.$$

9. Set t_{ff} equal to $t_{\text{gas-grain}}$: $\frac{5 \times 10^7\,\text{yr}}{\sqrt{n(H_2)}} = \frac{1.2 \times 10^{10}\,\text{yr}}{n_{H_2}}$.
This gives $n(H_2) = 5.8 \times 10^4$ cm^{-3}.

10 (a) $M = (4\pi/3)R^3 \times 1.36\rho_{H2} = (4\pi/3)[R(\text{pc})]^3(3.09 \times 10^{18}\,\text{cm})^3 \times 1.36 \times 2 \times 1.67 \times 10^{-24}\text{g} = 5.61 \times 10^{32}\text{g}[R(\text{pc})]^3 n(H_2(\text{cm}^{-3}))$.
1.36 is the 'helium correction': $1.36 \times 2 m_p = 4.54 \times 10^{-24}$g.
(b) Reformulate the above relation for M in solar masses:
$M(M_\odot) = 0.282[R(\text{pc})]^3 n(H_2)$.

Thus, $n(H_2) = (3.54M(M_\odot))/([R(pc)]^3) = 1.05 \times 10^3 \, cm^{-3}$.

(c) $\Delta V_{1/2} = \sqrt{(M(M_\odot)/(250R(pc))}$ yields $\Delta V_{1/2} = 16.3 \, km \, s^{-1}$.

(d) For 1000 GMCs, $M_{GMC} = 10^9 M_\odot$, so 33% of the mass of the ISM is in GMCs. The volume of the galaxy between 2 kpc and 8.5 kpc is

$V_g = \pi(R_2^2 - R_1^2)D = 1.15 \times 10^{66} \, cm^3 = 4.29 \times 10^{10} \, (pc)^3$. The volume of 1000 GMCs is $V_{GMCs} = 10^3 \times 4\pi/3R^3 = 1.41 \times 10^7 \, (pc)^3$. Thus, the GMCs occupy about 0.03% of the total volume.

(e) Column density: $N = Dn = 9.73 \times 10^{22} \, cm^2 \approx 10^{23} \, cm^2$.

Visual extinction: $A_v = 100^m$.

11. From the equation in 'Tools',

$$\frac{HD}{H_2} = \frac{D^+}{H^+} e^{\Delta E/kT}$$

with $\Delta E/kT = 500 \, K$. Then for $T=100 \, K$, the abundance of HD is $e^5 = 148$ times larger than the actual abundance. For the analogous reaction of CO and ^{13}CO, the energy difference is $\Delta E/kT = 35 \, K$, so at 100 K, the enrichment is 1.4. This is much less but still significant. There are effects that balance this enrichment, called 'fractionation', so that the measured and actual $\frac{^{13}CO}{^{12}CO}$ abundances from many different species agree rather well.

Index

© Springer International Publishing AG, part of Springer Nature 2018
T. L. Wilson, S. Hüttemeister, *Tools of Radio Astronomy – Problems
and Solutions*, Astronomy and Astrophysics Library,
https://doi.org/10.1007/978-3-319-90820-5

Printed in the United States
By Bookmasters